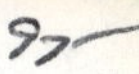

PROGRAM PLANNING AND EVALUATION FOR THE PUBLIC MANAGER

PROGRAM PLANNING AND EVALUATION FOR THE PUBLIC MANAGER

Ronald D. Sylvia
San Jose State University

Kenneth J. Meier
University of Wisconsin—Milwaukee

Elizabeth M. Gunn
University of Oklahoma

Prospect Heights, Illinois

For information about this book, write or call:

Waveland Press, Inc.
P.O. Box 400
Prospect Heights, Illinois 60070
(708) 634-0081

ISBN 0-88133-592-4

Printed in the United States of America

7 6 5 4 3 2

Introduction

Program evaluation grew out of a need for hard data on which to judge public programs, but progress toward general acceptance of evaluation science as a decision tool for public policy makers has been evolutionary. Unlike economists, whose arrival in the public policy arena was heralded by some as the coming of the solons, program evaluators were little noticed and seldom heralded. As far as some program officials were concerned, program evaluation represented another accountability clause that some politician insisted on including in the enabling legislation, and evaluators were another group of consultants whose presence had to be endured.

Unfortunately, this reception of early evaluation efforts was reflected in the quality of the studies produced. Because they did not understand evaluation, program officials often treated evaluators perfunctorily. For example, officials saw little purpose in answering queries about program goals, and they were not enthusiastic about explaining program operating procedures or detailing organization structures to academics. The evaluators, for their part, often found their designs and findings being methodologically critiqued by the same program staffs who had formerly expressed disinterest in the whole business.

Much of this confusion and conflict can be laid at the door of the program officials, who did not understand what evaluation was about and thus were not inclined to cooperate with the evaluator. But some of the problem can be traced to the methodological training that early evaluators received in such fields as parametric and nonparametric statistics and to research designs for the social sciences that emphasized the origins and development of the scientific method. This emphasis resulted in designs aimed at reaching objective conclusions about the relationship between two or more variables. Furthermore, because the goal of the individual researcher is to make a contribution to the body of scientific knowledge on a given topic, he or she is careful to ensure that others can replicate the research protocol and presumably confirm the conclusions.

But scientific objectivity does not impede the effectiveness of an evaluation. An overemphasis on professional detachment can obstruct an evaluation. The scientific method emphasizes the importance of maintaining professional detachment during research to ensure that the personal biases of the researcher are not reflected in the research design or the interpretation of the findings. Detachment also prevents the researcher from becoming so involved with the subjects that he or she cannot maintain a professional perspective. (This loss of perspective is sometimes referred to as "going native.") An overemphasis on professional detachment, however, can impair the evaluator's ability to aid line managers with program problems. An effective evaluator has the ability to balance the need for objectivity and the manager's need for solutions to problems. Accordingly, a bad evaluation is a greater danger than an evaluator "going native."

Evaluation researchers face other problems. Social scientists who engage in applied research risk professional disapproval by colleagues who tend to denigrate the application of research skills to applied problems as simply not being real social science. Some evaluation researchers can turn a deaf ear on such snobbery, but others attempt to modify evaluation designs in order to answer questions relevant to social theory rather than to program performance. Such designs are irrelevant to program managers. Administrators are most interested in why a program is not producing the desired result, not in what statistically does or does not cause what. When evaluators are too concerned with providing statistically rigorous designs, they are likely to gravitate to a measurement of the relationships between program outputs per unit of program input. Such a perspective does not lend itself to questions of organization structures, subunit productivity, or operating procedures, and it does not help the line manager solve problems that surface in the evaluation.

Such evaluations of overall program performance are defended on the grounds that they answer bottom-line questions of interest to legislative policy makers, but policy makers frequently do not have the time or inclination to dig through lengthy evaluation reports. At best, committee staffers may read the report, but even they may lack the necessary skills to interpret findings based on sophisticated designs. Ironically, the question that policy makers invariably raise in response to negative reports is "What is wrong with the program?" Narrow measures of input to output ratios cannot answer this question.

Evaluation research has progressed considerably since its inception. Evaluators have come to realize that if their data is to be meaningful it must be directed toward decision making. As such, the designs that are applied to evaluation have been expanded to include reviews of organization operating procedures, direct observations of organization performance, and interim reports of program progress which are submitted along with recommendations on how the organization can modify its activities in order to achieve desired outcomes.

As a symbol of the change in evaluator perspective, the expansion of designs is probably less significant than the increased willingness of evaluators to recommend operational changes. The latter reflects a problem-solving orientation and a willingness to forego professional detachment in the interest of achieving program outcomes. This change in evaluator perspective and the growing acceptance of evaluation as a management decision-making tool are reflected in the curricula of the burgeoning programs in public administration, many of which now require students to take one or more courses in program evaluation.

We would go a step further and have these classes include management-oriented problem-solving skills. Accordingly, we have chosen to define the field broadly. This text contains materials on systems analysis, program planning, and cost-benefit analysis, as well as on the subjects of applied research designs using outcome and process approaches to evaluation.

The text begins with a chapter on systems theory. Systems theory is presented in the belief that students must have a frame of reference for analyzing program structures if they are to be capable of making specific recommendations on program improvements. This chapter, however, makes the point that the benefits of having a framework for analyzing an organization may be offset if the analyst superimposes a systems framework where no real systems exist. Systemless organizations operate in a catch-as-catch-can manner when reacting to environmental demands and when nobody is sure what others do or why. In such circumstances, the systems framework may be most beneficial in helping organization actors see what their operations could be. When organizations do have established systems and procedures, a systems approach enables the evaluator to interrelate program outputs with the program units responsible for their accomplishment. The approach can also be a valuable tool in assessing how resources are allocated in the context of desired outcomes. Finally, systems theory can help the evaluator and program officials develop and implement organization adaptations that emerge in the course of an evaluation.

Chapter 2 deals with the planning function. We include planning in an evaluation text in the belief that the defining of organization goals, the design of systems for achieving those goals, and the creation of evaluation strategies for measuring program progress are interrelated. One cannot engage in program evaluation without some idea of what the organization is seeking to accomplish. Conversely, one cannot know if the program is proceeding as intended without at least a rudimentary review of program progress.

From an operational perspective, a careful analysis of the agency's mission and a systematic assessment of the linkages between the program and its environment are central to good management. Program managers must be able to anticipate changes in the environment of the program and make appropriate adaptations.

The chapter also introduces the student to the concepts of Program Evaluation and Review Techniques (PERT) and Critical Path Method (CPM). PERT/CPM have traditionally been associated with scientific and engineering programs such as aerospace research or the design and deployment of weapons systems. We believe that PERT/CPM are potentially useful for application to social programs. Therefore, the chapter concludes with a case study exercise designed to illustrate the social program potential of PERT/CPM.

Cost-benefit analysis is treated in Chapter 3. As a method of evaluation it is eminently applicable to the measurement of systems inputs to outputs and thus can be applied in conjunction with a systems approach. As a decision tool, however, cost-benefit analysis is most effective when used with other evaluation strategies aimed at measuring the effectiveness of organizational processes on a unit-by-unit basis.

Cost-benefit analysis is the form of evaluation that agencies are most often asked to perform. Frequently, a cost-benefit study is prescribed in the enabling legislation or as a standard part of organization operating procedures. The potential exists, however, for cost-benefit analysis to be applied when other evaluation formats would be more appropriate simply because the agency is comfortable with the cost-benefit approach. Chapter 3 is designed to acquaint students with how cost-benefit analysis works. The case studies at the end of the chapter seek to make the student aware of some of the pitfalls that accompany poorly planned or executed cost-benefit studies. For example, cost-benefit specialists could utilize assumptions that are either inappropriate or self-serving to the program, or both. By teaching students how to question cost-benefit studies, we seek to make them effective consumers of cost-benefit data rather than experts in the conduct of cost-benefit studies.

Chapter 4 deals with evaluation research designs and the threats to design validity and is based on the foundation laid by Donald T. Campbell and Julian C. Stanley. The chapter seeks to illuminate the problems of research design and validity using examples taken from program evaluation rather than from general social science. In addition, the chapter makes the point that program validity is as important as the internal and external validity issues relevant to social science in general. Social scientists emphasize design precision to ensure that the study measures what is intended (internal validity) and that others may replicate the methodology and findings (external validity). Program validity refers to the preeminent necessity of designing evaluations to answer questions relevant to program officials. Failure to do so may doom the evaluation findings to controversy.

Chapter 5, on conducting an outcome evaluation, begins by briefly addressing the problems of evaluator selection, staff resistance, and the various motives of persons involved in the evaluation process. Because it is important to achieve consensus among program officials with regard to the goals of the program, the chapter presents several approaches to building goal consensus, including the Delphi technique and group problem-solving techniques. Although consensus-building exercises are generally associated with process evaluations, they are presented here because they are equally applicable to outcome evaluations. Persons who engage in outcome evaluations often overlook the benefits of the group consensus approach.

The chapter then presents a model for designing and conducting an outcome evaluation. One feature of our approach is the inclusion of program elements in the model. Elements are delineated after goals are defined and before the indicators of program success are developed. Many outcome evaluations immediately move from the definition of program goals to the development of program success indicators. Our approach calls for the evaluator, in consultation with the program staff, first to identify program components and the organization units that are responsible for their implementation. The evaluation then will produce specific information on unit performances rather than merely generating aggregate assessments of program performance. This unit-by-unit information is necessary to managerial efforts to generate program adaptations based on evaluation findings.

Chapter 6 begins by distinguishing between the assumptions and methodologies of outcome and process evaluations, then presents an example of how to conduct a process evaluation in which the emphasis is on organization problem solving. The guiding principle of process methodology is a commitment to helping the program succeed. The process-oriented evaluator therefore mixes and matches evaluation designs in order to provide the organization with data it can use in making decisions. The methodologies presented here include the participant-observer approach, group interviews, and managerial audits. The chapter also points out how some of the methodologies associated with outcome evaluations can be applied in a process evaluation and emphasizes the importance of an evaluator's presenting solutions as well as identifying problems.

Chapter 6 also discusses how the evaluator helps implement the changes through various approaches. A system for managing change is necessary to ensure that changes are implemented as planned and to minimize the negative impacts of the changes. The chapter closes with a discussion of the importance of assessing the impacts of the changes on problems as well as on the organization generally. Ultimately, what a process orientation does is commit the evaluator to providing solutions as well as to identifying problems. The evaluator also takes partial responsibility for the success or failure of those solutions.

Chapter 7 treats issues related to establishing program evaluation as a resource to public managers. The chapter begins with a discussion of government attempts to deal with social problems and the contribution that systematic evaluations can make in these efforts. Such topics as the need for a national data base on factors related to poverty are discussed. Questions regarding the appropriate uses of evaluation technology are discussed next. Included in this discussion are the importance of design validity and the importance of formulating policy relevant questions in evaluation designs. The chapter then advocates the establishment of a national program of evaluation training for line managers. The chapter then discusses the development of a national network of evaluation clearinghouses so that designs can be shared with others who would be interested and so that program officials could quickly determine whether proposed program adaptations in their agencies had been attempted elsewhere and with what degree of success. Having data on the success of others could serve as a control for measuring one's own successes as well.

Finally, our combined experience in teaching program planning and evaluation has convinced us that the best way to teach students the techniques of evaluation design is to have them design evaluations. To this end, appropriate chapters are concluded with cases to help train students in applying the design strategies presented in the chapter. Some of the cases are modifications of evaluation, planning, or managerial problems found in the literature of public administration. Other cases are adapted from actual problems that the authors have researched or that were supplied by students from their experiences in government.

We would like to thank the following people for their helpful comments during the early stages of the manuscript: Richard Bingham, University of Wisconsin at Milwaukee; Richard Campbell, University of Georgia at Athens; Robert Durant, University of Georgia at Athens; Don Kettl, University of Virginia at Charlottesville; William Lyons, University of Tennessee at Knoxville; John Rehfuss, California State University at Sacramento; and Fred Springer, University of Missouri at St. Louis.

Ronald Sylvia
Kenneth Meier
Elizabeth Gunn

Contents

CHAPTER 1 SYSTEMS CONCEPTS 1

Systems Environments 2
- Laws 2
- Philosophy and Culture 2
- Economics 3
- Social Institutions 3
- Other Organizations 4
- Group Demands and Supports 4
- Human Resources 4
- Technology 4
- Material Resources 5

Inputs and Outputs 5
Goals 6
Feedback 6
Inside the System 6
- The Adaptive Subsystem 7
- The Managerial Subsystem 7
- The Technical Subsystem 8
- The Maintenance Subsystem 8
- The Support Subsystem 8
- The Accountability Subsystem 8
- The Marketing Subsystem 9

Suboptimization 9
From the General to the Specific: Applying Systems Theory 9
- Environment 10
- Inputs and Outputs 14
- Goals 15

The Advantages and Disadvantages of Systems Theory 16
For Further Reading 17

CHAPTER 2 PLANNING: THE OFTEN IGNORED FUNCTION 18

The Principles and Purposes of Planning 19
Steps in the Planning Process 20
- Recognize the Need to Plan 20
- Determine Program Goals 21
- Forecast the Future 22
- Set Priorities 24
- Develop Alternatives 25

Evaluate the Alternatives 25
Select the Optimal Alternative 25
Implement the Plan 26
Evaluate the Plan 27
Revise the Plan 27
Some Caveats About Planning 27
Difficulty in Attaining Consensus on Objectives 27
Oversimplification of Problems 28
Time Constraints 28
Information Constraints 28
Cognitive Nearsightedness 28
Political Constraints 29
Means Versus Ends 29
Values of Planners 29
The Planning Hierarchy 30
National Planning 30
Agency Planning 31
Program Planning 32
Unit Planning 32
Using PERT/CPM for Program Planning 32
The PERT Chart 33
The Critical Path 34
Altering PERT Charts 36
Monitoring a Program with the PERT Chart 37
Setting Up a PERT Chart 37
Other Issues in Planning: A Manager's Checklist 38
Program Definition 38
Structure Decisions 39
Personnel Decisions 40
Develop a PERT 40
Evaluation Decisions 40
Case Studies:
Wetlands, Waterfowl, and Waivering Resources 41
A PERT Application: The Case of Dusbow, Oklahoma 42
For Further Reading 45

CHAPTER 3 COST-BENEFIT ANALYSIS 47

Why Use Cost-Benefit Analysis? 48
Assumptions of Cost-Benefit Analysis 49
Impacts 50
Quantification 50
Individual Knowledge 50
Maximization of the Difference 51

The Techniques of Cost-Benefit Analysis 51
Identifying the Project 52
Listing the Impacts 52
Collecting Data 52
Classification of Data 53
Making Monetary Estimates 55
Pricing Techniques 55
Discounting the Values 62
Comparing the Costs and Benefits 65
The Net Present Value 65
Cutoff Period 66
Payback Period 66
Internal Rate of Return 66
Benefit/Cost Ratio 66
Equity 67
Sensitivity Analysis 67
Making Choices Based on Cost-Benefit Analysis 68
Identifying and Describing the Problem 68
Setting Up the Design 68
Collecting the Data 69
Analyzing the Data 69
Presenting the Results 69
The Pros and Cons of Cost-Benefit Analysis 69
Case Studies:
A Cost-Benefit Analysis of the National 55 mph Speed Limit 70
A Second Cost-Benefit Analysis of the National 55 mph Speed Limit 74
For Further Reading 78

CHAPTER 4 EVALUATION DESIGNS 79

Validity 79
Internal Validity 79
External Validity 80
Threats to Validity 81
History 81
Maturation 82
Testing 82
Instrumentation 83
Statistical Regression 84
Selection Bias 84
Experimental Mortality 85
Interaction of Selection with Other Variables 86

Four Threats to External Validity 86
Interactive Effects of Selection Bias and the Experimental Variable 86
Reactive or Interactive Effects of Pretesting 87
Reactive Effects of the Experimental Environment 87
Multiple-Treatment Interference 88
Design Selection 89
Experimental Designs 89
Nonexperimental Designs 95
The Participant-Observer Approach 100
Mixed Designs 101
For Further Reading 102

CHAPTER 5 CONDUCTING AN OUTCOME EVALUATION 103

Preplanning the Evaluation 104
Choosing between External and Internal Evaluation 104
The Evaluator and Management: Dealing with Hidden Agendas 106
Defining Program Goals 108
The Personal Interview 108
Consensus-Building Activities 109
The Importance of Goal Consensus 111
A Model for Outcome Evaluation 111
Legislative Intent and Program Goals 111
Program Elements 113
Proximate Indicators 113
Measures 113
Identifying and Assessing Program Outcomes 113
Outcome Valence 114
Applying the Model to a Community Health Program 114
Applying the Model to an Experimental Police Program 120
Case Studies:
A Hypothetical Case: Program Experimentation 123
A Hypothetical Case: Evaluating a Training Process 125
A Hypothetical Case: Migrant Workers 126
A Hypothetical Case: Teachers' Aides 127
A Hypothetical Case: Cutback Management 131
For Further Reading 134

CHAPTER 6 THE ART AND METHOD OF PROCESS EVALUATION 135

The Process-Outcome Dichotomy 135
Approaches to Process Evaluation 137

Problem Identification 138
- Surveys 138
- Group Interviews 139
- Supplemental Procedures 141

Presenting the Findings 142
Solution Development 144
Implementation 146
- The Task Force Approach 147
- Work Groups 147
- Scheduling Tasks and Setting Up a Reporting System 149

Feedback Evaluation 154
Overview of Process Evaluation 155
Case Studies:
- A Hypothetical Case: Balkinwalk, Virginia 155
- A Hypothetical Case: A Change of Command 157
- A Hypothetical Case: Troop B 159

For Further Reading 161

CHAPTER 7 PUBLIC POLICY AND SOCIAL PROBLEMS 162

The Socioeconomic Environment and Public Policy 163
Negative Effects of Public Policy 164
The Need for a Larger Data Base 165
Using Evaluation Technology Appropriately 166
- Problems of Design Validity 167
- Formulating Policy-Relevant Questions 167

Making Evaluations Relevant to Program Managers 171
- Evaluation of Program Structures and Procedures 171
- In-House Evaluation Units 172
- A National Evaluation Training Program 173
- A National System for Sharing Evaluation Information 174

Summary 175
For Further Reading 175

Glossary 177
Index 183

PROGRAM PLANNING AND EVALUATION FOR THE PUBLIC MANAGER

Chapter 1

SYSTEMS CONCEPTS

Both program planning and program evaluation are heavily influenced by systems theory. Systems theory is a conceptual framework for ordering one's thoughts about an organization or project. The terms *systems approach* and *systems analysis* refer to the applications of systems theory and its accompanying quantitative techniques. Other conceptual frameworks that could be used by the program manager are structural functionalism, causal modeling, and simulation. The popularity of systems theory, however, can be traced to three advantages: first, it strongly stresses the environment of a program; second, it permits an analyst to describe the interactions between a program and its environment; and third, it can be used heuristically without any of the sophisticated mathematics associated with program evaluation.

This chapter introduces the reader to the concepts and general philosophy of systems theory. It provides an introduction to many terms that will be used in the remaining chapters. The chapter begins with a discussion of systems environments and the elements that influence the processes of the organization. Then a description of a basic systems model is presented, including the concepts of goals, inputs, outputs, feedback loops, and the conversion process whereby inputs are transformed into policies and programs. The macrofunctions of organizations and the need for functionally specific subsystems are discussed next. The interface of organizations' environments and subsystems then are illustrated with the example of a spouse abuse treatment center. The chapter concludes with a summary of the advantages and disadvantages of systems theory as a framework of analysis and a method of program evaluation.

SYSTEMS ENVIRONMENTS

A simple model, illustrated in Figure 1-1, shows the basic elements of a system. A system receives inputs from its environment and processes them into outputs. Many systems theorists are content to ignore anything that transpires within the "black box" of the system. Analysts using this approach often concentrate on the ratio of outputs to inputs, cost-benefit analysis, input-output analysis, or cost-effectiveness analysis but, if program evaluation is intended to result in program adaptations, the interior cannot be ignored.

Systems analysis focuses much attention on the environment of a program. In general, the environment contains potential resources for and constraints on the program and places some value on the program itself or its outputs. Several specific aspects of a program's environment merit specification.

LAWS

Laws, both in the general sense of all laws and in the specific sense of the laws that authorize the program, affect the program operations. The criminal justice planner, for example, cannot solve the problem of overcrowded prisons by simply executing every fourth prisoner. In this case, general laws act as a constraint. A program evaluator examining a rehabilitation program must consider the requirement of federal law that half the people treated must be severely disabled. For rehabilitation programs, the specific law provides both program opportunities and program constraints.

PHILOSOPHY AND CULTURE

Philosophy refers to the public philosophy about specific government actions and may be termed *political culture.* For example, the United States operates under a philosophy that opposes government-operated enterprises, at least in the

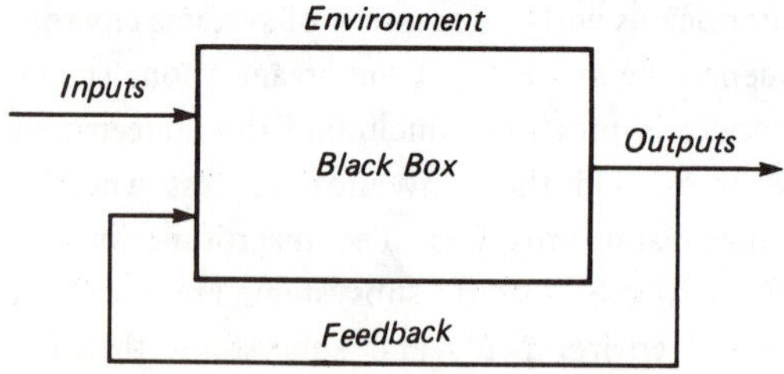

FIGURE 1-1
A simplified systems model

abstract.[1] For a program planner, this means that programs requiring government interference in the marketplace must be designed with some deception. As a result, transfer payments, for instance, are often called insurance programs. Private companies are often asked to run bankrupt railroads at cost plus a fee, even if the companies were themselves responsible for the bankruptcies, since public philosophy is less likely to support open nationalization. Program planners and evaluators should be able to specify any public values or general philosophical orientations that are likely to restrict program activities or present new opportunities.

ECONOMICS

Private-sector economic arrangements may affect program operations. Private-sector wages for airline pilots, for example, determine that the armed forces will be faced with a continual turnover of skilled pilots. Health programs must recruit doctors by offering incentives other than salary if government employers are to be competitive. Other economic arrangements dictate that research and development will be conducted by private-sector contractors rather than in public-sector laboratories. On the output side, regulators have recently been admonished to consider the impact of regulations on the marketplace. Finally, health and safety regulators have been especially subjected to cost-effectiveness criteria.

SOCIAL INSTITUTIONS

Other social institutions often constrain public programs. Although the impact of social institutions may be less now than in the past, the constraint is still present. A program planner seeking to design programs to prevent spouse abuse will be constrained by the American concept of the family and the societal support for family units. In many localities, the program planner cannot consider operating a program that encourages divorce as a solution, or attempt to use criminal prosecution as an alternative. Both options run against strong social institutions and would be opposed in many communities. Thus, the planner's options may be restricted to sheltering victims, counseling victims, and trying to provide viable alternative economic futures.

---

[1] In *The end of liberalism* (New York: Norton, 1969), Theodore Lowi contends that the United States operates under the public philosophy of interest-group liberalism. On the other hand, Daniel J. Elazar, in *American federalism: A view from the states* (New York: Thomas Y. Crowell, 1972), identifies three separate political cultures in the United States: the traditional culture, the moral culture, and a culture based on individualism.

OTHER ORGANIZATIONS

Other organizations often act as competitors for a given agency's resources or clientele. This must be considered in program planning and may be an advantage in program evaluation. The U.S. Postal Service, for example, must consider the actions of United Parcel Service, Federal Express, and other similar firms when contemplating alternative systems of parcel post. These private-sector organizations limit Postal Service options. In the evaluation of programs, on the other hand, organizations that offer comparable programs can provide direct comparisons to the program to be evaluated. When such comparisons are made, however, care must be taken to guarantee that the programs *are comparable.*

GROUP DEMANDS AND SUPPORTS

The demands and preferences of program clientele should be carefully considered in both evaluation and design. Not only do these demands act as constraints or resources, depending on their nature, but they often can be used to generate program goals and objectives. The impact of groups other than the clientele should also be considered. These groups may have information of value to the evaluator/planner. They include such groups as agency employees, professionals with program expertise, legislative staffs, audit personnel, and other outside interest groups.

HUMAN RESOURCES

The availability of human resources often affects program planning and should affect program evaluation. For example, a health care delivery system plan for a rural area is limited by the number of medical personnel willing to practice in that area. Paramedics and midwives may be the only viable options. On the other hand, surpluses of social workers permit a welfare agency to hire more workers at less cost and thus consider alternatives that are labor intensive. Evaluators, especially process evaluators, need to be aware of any human-resource problems that may affect program performance.

TECHNOLOGY

The state of technology directly affects program-planning efforts. In order to plan effectively, the manager must forecast technological advances. Future U.S. Postal Service programs, for example, are dependent on microwave transmission capabilities. Energy programs that involve solar, shale oil, and other new energy forms are dependent on the state of technology. Some legislation, such as the National Environmental Protection Act, requires that the administrators consider possible new technologies when planning programs. A program evaluator also must be concerned with technology. Process evaluators, for example, need to know about alternative technologies for processing information or for producing program outputs.

MATERIAL RESOURCES

Material resources include physical facilities, energy, and hardware as well as money. In an era when many governments with stable or declining resources are facing increasing demands, the availability of resources is an essential element in program planning. Some government agencies need to plan to conduct programs with fewer material resources. The availability of material resources plays a role in evaluation; some government programs are doomed to fail because they have inadequate or even inappropriate resources.

INPUTS AND OUTPUTS

Systems theory holds that organizations and programs receive inputs from the environment and provide outputs valued by the environment. Inputs can be broadly categorized as either demands or supports. In program terms, we call them constraints and opportunities. Each element of the environment may generate constraints or opportunities to a given program. In general, program planning and evaluation require a full inventory of program inputs. More specifically, these inputs form the basis for several types of input-output analysis, including cost-benefit and cost-effectiveness analysis.

If the outputs are adequately valued by the environment, the organization can procure sufficient inputs to survive. There are several categories of program outputs: a program may provide goods (missile systems), services (employment counseling), regulations, adjudications, or support for other programs (recruitment).

For output evaluation and output-oriented program planning, the concept of outputs is essential. Outputs may be either direct—outputs the program was designed to achieve (a drug abuse program lowering drug abuse) or second-order outputs (reduced street crime resulting from a successful drug treatment program).

Many systems theorists need only these concepts to engage in planning or evaluation. For them, a successful program is defined as one that maximizes output based on specified inputs, subject to environmental constraints. The general form of linear programming, as well as input-output analysis, is based on this definition.

GOALS

The concept of program goals is central to systems theory. Systems are presumed to be pursuing goals. Most general-systems theorists argue that the goal of a system is to survive, to avoid the pressures toward entropy. This statement, however, is of little use to the program evaluator. Both program planning and program evaluation are more normative; programs should exist because they fulfill some worthwhile goal.

Program analysts therefore focus on more limited goals. What is this program intended to achieve? What processes are essential to an acceptable program output? How can the organization be designed to streamline processing? Classical program planning and classical program evaluation both require that program goals be expressly stated. A program with a large input-output ratio that does not attain its goals cannot be considered a success.

FEEDBACK

One special type of program input is feedback. The environment receives outputs and, if the correct mechanisms exist, feeds back information about its reaction to the outputs. In one sense, program evaluation is a type of feedback generated about a program. Feedback can also be generated by reports, audits, clientele reactions, responses of other social institutions, and so on.

Feedback permits an organization or program to be adaptive. If an acceptable range of outputs is forthcoming, nothing need be done. Negative feedback tells the organization to adjust outputs. All program plans need to contain feedback systems so that program errors and limitations can be corrected.

INSIDE THE SYSTEM

Although many systems theorists ignore what goes on within the system (inside the black box), others believe that every system is composed of subsystems. For the effective functioning of a system (or program), each subsystem also must function effectively. One view of internal systems specifies six subsystems essential to the operation of any program or organization—the adaptive, managerial, technical, maintenance, support, and accountability subsystems. The interrelationships of these subsystems are illustrated in Figure 1-2. A seventh subsystem, the marketing subsystem, is found in some programs too.

THE ADAPTIVE SUBSYSTEM

The adaptive subsystem is concerned with anticipating the future and preparing for it. This subsystem deals with strategic and operational planning. Programs lacking an adaptive subsystem may continue indefinitely, even if no one benefits from their existence other than program employees. The Subversive Activities Control Board of the 1960s is a prime example of a system's failure to adapt by discontinuing obsolete subsystem activities.

THE MANAGERIAL SUBSYSTEM

The managerial subsystem defines the goals for the program and organizes the program around those goals. Chester Barnard, an early managerial theorist, believed that upper management's function was to determine structural arrangements. Once these structural decisions were made, the remainder of the program's decisions could be made by middle- and lower-level managers. Programs

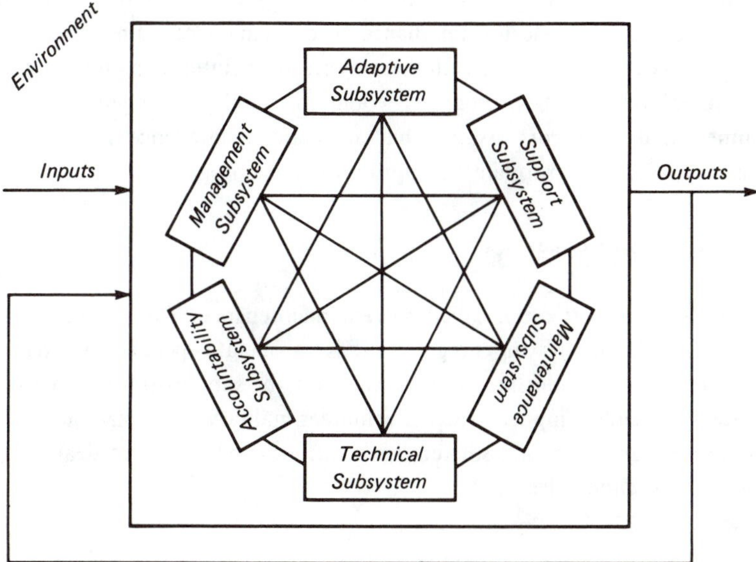

FIGURE 1-2
Subsystems of a program

can survive with an ineffective managerial subsystem, but they cannot survive long without any managerial subsystem at all. The importance of the managerial subsystem increases in a turbulent environment. When management must constantly adapt a program to changing environmental demands, ineffective management may be fatal to the program. On the other hand, if the environment is stable, programs can flourish almost without top management once decisions have been routinized.

THE TECHNICAL SUBSYSTEM

The technical subsystem is the mainline production apparatus of a program that produces the good or the process. If the program creates tanks, the technical subsystem would be said to produce a good, namely tanks. If the program produces a service, such as employment counseling, then the technical subsystem produces a process. Although some systems theorists believe that all technical-subsystem problems can be traced to inadequacies in other subsystems, most theorists believe that weak technical subsystems can be corrected by focusing on those technical subsystems.

THE MAINTENANCE SUBSYSTEM

The maintenance subsystem performs the procedures necessary for program continuation. Routine physical maintenance is one example; maintenance of employee records is another. Without these types of maintenance and the enforcement of job-related rules, programs would soon collapse. In general, program planners and evaluators assume that adequate maintenance subsystems exist. This may be an inappropriate assumption for some programs.

THE SUPPORT SUBSYSTEM

The support subsystem is often called pattern maintenance. The support subsystem procures the inputs the program needs to operate—people, materials, technology. Programs can decline in effectiveness if the support subsystem cannot provide the needed inputs. The all-volunteer military, for example, has trouble recruiting and retaining soldiers with sufficient education to deal with modern weapons technologies.

THE ACCOUNTABILITY SUBSYSTEM

The accountability subsystem serves as the internal management information system. This subsystem designates the reports, audits, and evaluations that must be performed for use by internal management. Not only can accountability subsystems be too weak, they can also be too strong. In rehabilitation agencies with elaborate reporting systems and production quotas, some personnel may compete for easy "rehabs" and avoid the more difficult cases. This may be rational for the individual counselor, given the accountability system, but it is counterproductive for the program.

THE MARKETING SUBSYSTEM

Some programs have a marketing subsystem that "sells" the program's outputs to the environment. Other programs are basically passive. Food stamp programs provide examples of each. Some states operate aggressive outreach programs to convince qualified people to use food stamps. Other states have passive programs. If you come in, food stamps are dispensed, but no one is encouraged to come to the office. Since, in the public sector, programs can operate without marketing subsystems, the marketing subsystem is not counted as an essential subsystem.

SUBOPTIMIZATION

All the subsystems of a program are interrelated and each affects the others. The managerial subsystem, for example, provides goals and structure for the technical subsystem. In turn, the accountability subsystem provides information for the managerial subsystem. Because all subsystems are interrelated, suboptimization may be a problem. Suboptimization occurs when one subsystem attempts to maximize its goals without any concern for its impact on other subsystems. When this subsystem optimization is harmful to the overall system, it is called suboptimization. For example, an accountability subsystem that maximizes reports, audits, and controls would prevent overall optimization because many workers would be filling out forms rather than producing program outputs.

FROM THE GENERAL TO THE SPECIFIC: APPLYING SYSTEMS THEORY

The preceding discussion of systems theory is fairly abstract. To illustrate how each of these elements can be used in program planning or program evaluation, an extended example is in order. We will use a domestic violence program for that example.

The Spouse Abuse Center is a shelter program for women who are victims of domestic violence. The center, a large house in a residential neighborhood, provides rooms for women and their children and counseling and assistance during the victims' transitions to a nonviolent environment. It was founded by a local feminist organization in a university city and is run by a private, nonprofit organization, financed with public-sector funds.

We will assume that the Spouse Abuse Center wants to be evaluated. The systems framework will be used in an effort to illustrate how each element in the systems framework has specific application in a program evaluation.

ENVIRONMENT

First we discuss nine elements in the center's environment that have direct impacts on the operation of the program.

Laws. The evaluation must initially consider the law that created the agency or program. This may be a federal law, a state statute, a local ordinance, or some combination of the above. Since laws are often vague, they must be supplemented by a reading of administrative regulations and the legislative history. Administrative regulations are fairly easy to find at the federal level since they must be published in the *Federal Register.* State and local governments have varying procedures and therefore require greater effort to find. The legislative history of a law consists of all supporting materials that precede and follow the legislation. Committee hearings, floor debates, and final reports often contain very specific discussions of program goals and agency procedures.

For many programs, court decisions must also be considered. According to the U.S. Supreme Court, the Occupational Safety and Health Administration, for example, is not required to consider costs when promulgating rules. The court thereby mandated to OSHA authority that may exceed the intent of Congress in creating the agency.

In the Spouse Abuse Center example, the legal environment is simple. Since the center is a private organization, there are no enabling laws to consider in the evaluation. This does not mean the center has no legal environment. Several laws are relevant to the center, including the state statutes on assault, laws on residency requirements to receive welfare payments (for discharging clients), legal procedures to prevent harassment, and so on.

Philosophy and Culture. Daniel Elazar identifies three predominant political cultures in the United States–traditional, individual, and moral.[2] Traditional

[2] Daniel J. Elazar, *American federalism: A view from the states* (New York: Thomas Y. Crowell, 1972).

political culture sees government's role as the maintenance of social order with the government controlled by an elite. Individual political culture sees government as passive, facilitating the actions of the marketplace and little else. The individual is seen as the source of all value. Moral political culture sanctions an active government seeking to remedy the political, social, and economic ills of the society.

The Spouse Abuse Center is located in a community with a mixture of traditional and individual political cultures. This means that little cultural support exists for an activist social-service agency. In fact, citizens would be skeptical of the need for such an institution.

Economics. The economic variables of initial concern to the program evaluator are the general indicators of the economy–the inflation rate and the unemployment rate. High inflation rates mean that an agency's operating costs will increase in future years. High unemployment rates mean that government tax collections will decline and, therefore, that agency budgets will be static. Nationwide estimates of future inflation and unemployment are obtainable from several sources–the Congressional Budget Office, the Office of Management and Budget, and private forecasting firms. Estimates for local areas also are obtainable, but these are of questionable accuracy.

For the Spouse Abuse Center, the general forecast is positive. Unemployment rates are very low, while inflation is moderate compared with the nation as a whole. As a result, state funds for the center are reasonably secure.

The general economic indicators also may have a direct impact on programs. For example, incidents of spouse abuse increase during economic hard times. Economic frustrations of the husband often manifest themselves in aggression in the home. The economic outlook, therefore, can be used to project center work loads.

Subsections of the economy also must be considered. Projected price increases for food, housing, labor, and manufactured goods may be relevant. The center owns its shelter, so housing costs only affect possible expansion. Food and medical costs, however, are major considerations and an evaluation must be concerned with controlling these costs.

The other major economic variable that the center must consider is the cost of its personnel. If some personnel inputs are prohibitively expensive, alternatives must be examined. The center is fortunate because it is located in a university community. Many of the personnel it needs–social workers, counselors–are readily available in large numbers, so the center's personnel costs can be expected to be low.

Social Institutions. The evaluation should ask whether the program to be analyzed affects or is affected by other social institutions in any major way.

The analyst needs to consider the family, religion, local businesses, the educational system, and other nongovernment institutions. (Government institutions are included in the next section.)

The Spouse Abuse Center runs directly into two dominant institutions: the family and religion. Two of every three center clients divorce their husbands. Members of the staff have been known to joke that the center's goal is to destroy marriages. Whether this view of the center's goals is correct or not, the center's role as an alternative to remaining in a violent situation ensures that the center will have some hostile publics. This fact is underscored by the predominant local religion. The center is located in a heavily fundamentalist community in which religious beliefs support the idea that the man is the head of the household and the woman is subordinate to him, no matter what. One religious woman stated to the center staff, "My husband only beats me when I deserve it." Such institutional constraints must be considered by any evaluator or program planner who is analyzing community outreach programs.

Other Organizations. The analyst needs to be creative in identifying other organizations that affect the agency's operations. At a minimum, the analyst must consider the funding institutions and any functional rivals (others that perform the same service).

The Spouse Abuse Center initially received funds from the federal government (CETA and VISTA funds), from the state government (mental health funds and community development funds), from charitable organizations (United Way), and from private individuals. With the advent of the Reagan administration, CETA and VISTA funds dried up; this mandated program planning for alternative funding sources. State funds appear to be reasonably stable, but there are no long-run guarantees because the center is a private organization, not a government agency. The United Way is concerned about accountability and has been asking the center to evaluate its programs. Each of these organizations therefore affects program planning and must be considered in any evaluation. For example, the mental health agency will establish evaluation criteria that are different from those of the community development agency and the United Way.

The Spouse Abuse Center has no functional rivals in its service area. Several other centers exist in the state, however, and these could be used for comparisons in the evaluation. Other organizations the center interacts with include the police, the district attorney, local welfare offices, local hospitals, and the local neighborhood association.

Group Demands and Supports. Numerous other outside groups must be considered in an evaluation. A general list of these groups includes clients, potential clients, contributors, agency supporters, agency critics, agency employees, and any other groups not considered elsewhere. The agency staff is often the best source for a list of such groups.

For the Spouse Abuse Center, all these groups are relevant. Clients tend to lack political power and are not likely to publicly identify themselves to the evaluator. As a result, client perceptions of the quality of service may not be obtainable. People who contribute funds to the center are likely to be program advocates, although they may also be a source of ideas about how the center should be run. Supporters and critics need to be analyzed in terms of their numbers and the intensity of their interest in the center. Both can provide the evaluator with information that can be used in an evaluation.

Human Resources. Human resource considerations go beyond the cost of these inputs. The evaluator needs to consider the motivations, the training, the skills, and any strengths and weaknesses of the human resources. Proposals for change that are beyond the capabilities of the present staff or inconsistent with their view of the organization's role are likely to fail.

The staff of the Spouse Abuse Center present some unique characteristics. They are all members of the "helping" professions–social workers, counselors, and so on. The attitudes of these professionals are overlaid by strong feminist beliefs. The center grew out of a local feminist organization and retains the attitudes associated with the feminist movement. Managerial skills, on the other hand, are fairly scarce. None of the staff has any training in public or business administration. In fact, the staff members express some hostility toward such management techniques as evaluation, performance appraisal, and accounting. The center is run by consensus decision making and has a fluid organizational structure.

The implications for program evaluation are clear. The staff will not be eager to assist any evaluation effort; they will cite the need to deliver services. Furthermore, attempts at process evaluations of agency procedures will be frustrated because the evaluator is an outsider. Attempts to program plan for changes in management style will be resisted or subverted. On the positive side, staff members are zealous and are highly committed to agency programs: they work long hours for low salaries because they believe in the program goals.

Technology. In the case of the Spouse Abuse Center, the technology is a process involving the provision of shelter and nurturing to victims of spouse abuse, as well as individual and group counseling services. A strictly output-oriented evaluation might ignore the technology subsystems (client counseling, providing a safe refuge) that are engaged in service delivery because of a lack of concrete measures of the processes. A process-oriented evaluation, on the other hand, would focus on these processes in an attempt to upgrade the quality of services in the center.

Because staff members of the Spouse Abuse Center come from the helping professions and resist the imposition of management systems, the evaluator would do well to emphasize a cooperative mode involving self-assessment by the staff. The evaluation might begin with a group session in which program professionals would discuss service delivery problems in the context of program goals. Second, the staff might be asked to review the relative emphasis that the agency does and should place on various services. For example, the center provides housing in a safe haven and counseling to individual victims in one-to-one and group formats, as well as services to clients who call the center for advice and counseling but do not need shelter. Finally, the center engages in outreach activities aimed at a target audience of current potential victims and conducts a program to increase public awareness.

Once the staff has concluded its assessment of what they currently do vis-á-vis the intended goals of the agency, the evaluator can assist them in the development of systematic programs to bring services into line with intent. Once program adaptations are complete, the evaluation can again assist the staff in an assessment of the changes. Or the evaluator can engage in more systematic measurement of the changes involving client surveys, cost-benefit studies, and the like. (A fuller discussion of the process approach is presented in Chapter 6.)

Material Resources. The material resources of an agency constitute a major constraint on all program activities. Physical facilities, equipment, and money are the main categories of material resources. The Spouse Abuse Center owns one house with a capacity of thirty persons. Normal operating capacity is ten women and twenty children. The building has been full since it opened. The house is fifty years old, is located in a transitional neighborhood, and was renovated before the center began operation. The house is valued at $120,000 with a $40,000 mortgage outstanding. Material resources associated with the house include a complete set of furnishings donated by private individuals. Equipment resources are nonexistent.

Money resources of the center are $68,000 from the state Department of Mental Health, $75,000 from the state Community Development Office, $18,000 from United Way, and $3,000 in private contributions. The total monetary resources of the center are $164,000.

INPUTS AND OUTPUTS

There are two general types of evaluation: outcome evaluations and process evaluations (see Chapters 5 and 6). Outcome evaluations contrast inputs with outputs to determine how well a program works. Process evaluations analyze the processes an agency uses by comparing them with some ideal. In this case we can apply outcome measures to supplement the process approach discussed earlier.

Using the preceding description of the environment, an enumeration of agency inputs should be compiled. For the Spouse Abuse Center, there are the following inputs:

Supports	*Demands*
Center budget of $164,000	*X* number of abused spouses
House and furnishings	$40,000 mortgage
Highly motivated staff	Capacity constraint of thirty
Private contributions	Lack of management skills
Low cost of personnel inputs	Powerless clients
Healthy economy in the area	Restraints placed by funding sources
Laws on spouse abuse	Neighborhood resistance
	Family and religion restraints
	Political culture

Outputs are sometimes easy to list when the agency produces things, for example, missiles, paved roads, and so on. Agencies that are service oriented, however, have less-certain output indicators. Such is the case with the Spouse Abuse Center. The number of women sheltered can easily be determined. In the philosophy of some staff members, this is the sole output of the center. They feel the center exists to provide a safe place for women to contemplate their options for the future. Other staff members point to employment and psychological counseling offered by the center. Still other staff members believe that a client is served only when she is permanently removed from a violent situation. Cases in which the woman returned to her home are not counted by these staff.

Input-output ratios can be calculated for a variety of measures (the cost of one day's shelter for one woman can be found) or the input-output analysis can be descriptive without any quantitative analysis. The type of input-output analysis depends on the organization, the evaluator, and the organization's goals.

GOALS

The stated goals of the Spouse Abuse Center are "to provide a secure place for a battered spouse to consider the alternatives open to her and to provide guidance and counseling to help the woman implement any decision she makes." The creative evaluator can translate this statement into several operational indicators of the center's success:

- Security of the shelter
- Client perception of alternatives
- Client evaluation of counseling
- Number of cases of abused women in the community versus number served by the shelter
- Costs of alternative shelters
- Visibility of the center to women in the community

THE ADVANTAGES AND DISADVANTAGES OF SYSTEMS THEORY

Systems theory offers several advantages if it is used in program planning and evaluation. First, both program planning and program evaluation use many of the same terms that systems theory defines: goals, inputs, feedback, environmental constraints, and suboptimization.

Second, systems theory illustrates the interaction of the subsystems and the interaction between the system and its environment. Good program plans and good evaluations need to recognize these interactions and adjust the analysis to them. Systems theory provides a framework that forces the manager to think about possible interactions.

Third, systems theory illustrates the constraints on managerial actions. Environmental restrictions may limit program-planning options. Subsystem limitations may prevent overall program success.

Fourth, systems theory shows the minimum number of functions that must be performed in a program. As such, systems theory can be used as a checklist in program analysis. For example, have we fully specified all inputs, or has the support subsystem been examined?

Despite the utility of systems theory in program analysis, the analyst must recognize that there are several limitations. First, systems theory does not yield clear-cut prescriptions. The theory tells analysts that a support subsystem is essential to a program, but it does not tell them the best way to design a support subsystem or how this subsystem must be linked to other subsystems. Systems theory is a conceptual framework; it is not a precise analytical technique.

Second, the use of systems theory may artificially force all programs to fit within the framework of systems theory. We do not know that all programs in fact operate as systems. The analyst faces the danger of incorrectly analyzing a program because the program has been inappropriately forced into a systems framework.

Finally, several of the systems concepts are not precise. In the real world, it is often difficult to distinguish between a system and its environment. In addition, some programs have a single subsystem that performs two or more of the subsystem functions, which makes it difficult to distinguish between subsystems. Using the systems framework may cause the analyst to be too precise.

The foregoing problems notwithstanding, systems analysis can be a useful tool for the program planner and evaluator, provided it is not asked to exceed its analytic and prescriptive limits.

FOR FURTHER READING

Emery, F. E., ed. *Systems thinking.* New York: Penguin, 1969.

Haberstroth, C. J. "Organizational Design and Systems Analysis." In *Handbook of organizations,* edited by J. March, pp. 1171–1212. Chicago: Rand McNally, 1965.

Miles, R. F., Jr., ed. *Systems concepts.* New York: Wiley, 1973.

VanGigch, J. D. *Applied general systems theory.* 2nd ed. New York: Harper & Row, 1978.

White, M. J., et al. *Managing public systems: Analytic techniques for public administration.* North Scituate, Mass.: Duxbury Press, 1980.

Wright, C., and Tate, M. D. *Economics and systems analysis: Introduction for public managers.* Reading, Mass.: Addison-Wesley, 1973.

Chapter 2

PLANNING: THE OFTEN IGNORED FUNCTION

A frequently heard management axiom purports: "A manager who has no time to plan has no time to live." Managers beset by the press of day-to-day events protest, "It is hard to devise a plan for draining the swamp when you are up to your fanny in alligators," to which the planners reply, "You should not go into the swamp without a contingency plan for alligators." In our view, the importance of planning cannot be overemphasized.

The percentage of a manager's time consumed by planning increases as one advances up the hierarchy. Unfortunately, some managers are never quite able to change their perceptions of their roles from responsibility for carrying out the mission to that of accepting responsibility for mission definition. The purpose of this chapter is to provide the manager with a rudimentary knowledge of the techniques of program planning.

Inclusion of a planning discussion in a text on program evaluation may seem strange to those who believe that planning should be discussed as part and parcel of general public management texts. We believe, however, that it is much easier to evaluate a program that is carefully planned than a program that has merely evolved. We also believe that, if program evaluation is ever to become an effective tool of management, evaluation findings must be used for clearly planned, systematic program adaptations.

The interrelationship of planning and evaluation is obvious. Program impacts are difficult to measure in the absence of clearly defined goals and accountability for program successes and failures is difficult to affix in the absence of clear-cut delineations of program responsibility and authority. These responsibility/authority lines, in turn, are the product of program planning. Conversely, an important step in program planning and effective implementation is a periodic evaluation of program progress.

The discussion begins with an overview of the planning function and its generally applicable principles. Various types of planning are then discussed, including long-range planning, the importance of assessing changes in the organization's environment, and the principles of forecasting. Next, program planning is examined, including the techniques of PERT/CPM. Finally, the discussion focuses on the planning process and the kinds of organization structures that are appropriate for carrying out the plan.

THE PRINCIPLES AND PURPOSES OF PLANNING

Planning fulfills several important functions for the organization. First, a plan defines the activities and direction of activities for those in the organization. A plan tells the workers where the agency is going and often when it is going there. This allows employees to discern how their individual efforts fit into the agency's overall activities.

Second, a good plan establishes criteria that the manager can use to make decisions. For example, plans contain goals and objectives for the organization that can be used by line managers in program designs and resource allocations.

Third, a well-constructed plan permits evaluation. As later chapters of this book discuss, evaluation is not possible without comparisons to some standard, even if the standard is limited to the notion that benefits should exceed costs. Some plans even include projected measures of program success. At the most basic level, it is possible to make crude evaluations by comparing program results with plan objectives.

Fourth, planning limits the quality and quantity of the control information that is gathered. In areas with long histories of quantification, such as rehabilitation administration, many times more data are gathered than can possibly be used. The result is a communication overload, whereby valuable information may be lost in an ocean of other data. If a plan clearly defines program objectives, and if these program objectives are translated into measurable indicators of program effectiveness, the amount of data needed for control is limited. The plan also can be used to design the agency's management information system, including data analysis capacity, inventory control, automated record systems, and human-resource planning.

Fifth, effective planning can minimize costs by smoothing work-load fluctuations. For agencies with cyclical work loads, such as the Census Bureau or a budget office, planning can be used to reallocate nonessential tasks away from peak demand times. For agencies organized around projects (such as NASA), planning can permit smoother transitions between projects. For all agencies, planning around work-load fluctuations can coordinate the divisions within an agency and avoid some suboptimization.

Sixth, planning, especially using such techniques as PERT and CPM (discussed later in this chapter), permits the agency to schedule. Comprehensive plans permit scheduling of tasks, personnel, facilities, outside contracts, and monetary resources.

Despite the numerous purposes of planning, the consensus is that public organizations do little planning. The planning that is done is often done only because it is required by law. A great deal of effort may be expended for a plan that gathers dust. The general lack of effective planning has been explained by James March and Herbert Simon in terms of Gresham's Law of Planning: Just as cheap money drives out good money, the pressures of daily activities drive out the opportunities for planning.[1] Unfortunately for the organization, many problems become crises that prevent planning simply because the organization failed to plan in the first place.

STEPS IN THE PLANNING PROCESS

The steps in the planning process are presented here in general terms, without the substantive program in which the planning takes place. We recognize, however, that planning realistically cannot be separated from substantive knowledge when the planning takes place. Without detailed substantive knowledge of the program and its content, the planner can contribute little to the organization. Substantive program knowledge is indispensable at all stages of the planning process.

RECOGNIZE THE NEED TO PLAN

The first step in the planning process is to recognize that planning can make a valuable contribution to program operations. In some cases, legislative requirements force agencies to plan: the federal government is likely to require that

[1] James G. March and Herbert A. Simon, *Organizations* (New York: Wiley, 1958), pp. 173-211.

state agencies submit plans before they can receive federal money. Often, however, such plans are merely window dressing because detailed program planning is rarely required. Nonetheless, program managers who take planning seriously reduce the day-to-day problems of program management and effectively implement public policy.

DETERMINE PROGRAM GOALS

Planning cannot be undertaken without clearly defined goals because program goals become decision criteria as the process evolves. A good set of program goals includes service goals, effectiveness goals, and efficiency goals. Questions relevant to the planner include: What service is to be provided and to whom is it to be delivered? How will we know if we have delivered an effective program? What are the minimal standards for success? Under what criteria can the program be judged for efficiency? What should a minimally effective program cost? Are there comparable programs that can be used to measure efficiency?

Program goals can occasionally be found in authorizing legislation. Where they exist, these legislative goals are a useful starting point. However, a program may not have legislative goals for two reasons. First, in many cases the legislature may provide an agency with only a broad, general mandate. Under such circumstances, the agency must devise both the program and the goals. Second, goals may be missing from legislation because the legislature could not agree on the goals. In an incremental political system such as ours, goals often cause conflict, so negotiation is done on programs rather than on goals. The result is programs without clear goals; the agency must supply them.

Goals need to be stated clearly and concisely, with enough information to serve as decision criteria. An example of a weak goal statement would be the Agricultural Research Service defining its goals "as doing agricultural research." A much better goal statement might read: "Improving the quality and quantity of the food supply in the United States by funding and conducting applied research on animal and plant productivity." This goal statement identifies the overall criteria (improving the quantity and quality of the food supply in the United States), the means (funding and conducting applied research), and the target (animal and plant productivity).

The manager should be aware that setting program goals for planning also has positive ramifications for job performance. Studies of the relationship between clearly specified goals and job performance generally show positive results. Performance tends to be higher when goals are specified than when they are not. The more difficult the goals, the greater the performance, provided the workers accept the goals. When conflicting goals exist, workers try to maximize the clearer goals and ignore the ambiguous ones. In short, a positive second-order consequence of setting program objectives should be improved employee performance.

FORECAST THE FUTURE

Because planning is a future-oriented activity, effective planning requires the manager to forecast future program environments. In general, forecasting has five steps. The *first step* is to describe the organization and its current environment. This is done by asking such questions as: Who are we? What is it that we do? How do we accomplish the mission and for whom is it done? Program-specific inputs, outputs, and environmental constraints should be identified. Identifying agency enemies, as well as friends and potential friends, is a prerequisite to developing an effective plan. Similarly, an organization's self-assessment of current organization patterns must be done in the context of its environment. Failure to recognize the need for more fuel-efficient automobiles, for example, put American manufacturers at a competitive disadvantage with foreign manufacturers. Changes in the environment of public organization occur even more frequently than in the private sector. For example, an incoming political administration may move to cut an agency's budget substantially. When this happens, knowing which elements of the program are essential is critical. Moreover, fending off budgetary attacks requires the aid of friends in the environment. Agencies that cannot mobilize environmental resources will sustain greater damage in austere times than agencies that cultivate linkages with actors in the political environment.

The *second step* in forecasting is to project changes in technology that affect the agency. A U.S. Postal Service forecast, for example, would include forecasts of communications technology such as satellites, microwaves, computer terminals, and other methods for transmitting information. The feasibility of different technologies affects both the demand for traditional mail service and the need to adapt to new service demands.

The *third step* involves forecasting changes in agency clientele. The Department of Agriculture, for example, needs to know that its clientele will be fewer in number but larger in economic concentration.

The *fourth step* in forecasting is to identify future opportunities and obstacles. These may be gleaned from steps one through three or result from a separate forecasting activity. For example, the Postal Service would consider its high fixed-labor costs to be an obstacle and new information-processing techniques to be opportunities.

In the *fifth step,* any and all program-relevant forecasts should be made. If one is concerned with defense planning, for example, a forecast of Soviet activities is essential. An effective defense-planning system also would develop a set of contingency plans to deal with such eventualities as the overthrow of a friendly regime around which strategic plans for an entire region had been constructed.

Regardless of how careful an agency is to forecast all contingencies, perfect knowledge is generally not possible and, in most cases, is not desirable because of the costs involved. Determining the number of program-relevant forecasts that are worthwhile is a managerial decision that is highly variable. The prudent manager, however, realizes that knowledge is power and that time spent planning is later saved through timely and effective program adaptations.

Forecasting Methods. Several forecasting methods can be used by the manager. The most common technique is known as ***genius forecasting,*** in which one person or a group of persons speculate about a program's future environment. Genius forecasting is based on the knowledge, intuition, and hunches of the individual genius. Genius forecasting has the obvious limitation that the forecast is only as good as the genius doing the forecasting. Sometimes an agency will use several genius forecasters so that major errors will be checked. Since genius forecasts may be vastly different, a mechanism is needed to reconcile divergent forecasts.

The most common way of generating consensus on forecasts is the ***Delphi technique.***[2] Under the Delphi procedure, several people forecast the agency's future environment independently of each other. All forecasts are then tabulated, and the summaries are fed back, without identification, to the other forecasters. Each forecaster then revises his or her forecasts, taking into consideration the logic, reasoning, and forecasts of the other forecasters. With the anonymity of the process, forecasters are in theory free to change their forecasts based on the Delphi feedback. Compilations are again recirculated to the forecasters for another round. The process continues until a consensus is reached.

Delphi is not a forecasting technique; it is a consensus technique. Delphi can be used to build consensus on policies, goals, and programs as well as on forecasts. As a consensus technique, it has some limitations. First, the consensus forecast may not be the best forecast but one that is limited by the common perceptions of the forecasters. Second, truly creative forecasts may be discouraged in the iterative process.

Trend extrapolation is a common mathematical method of forecasting the future. Trend extrapolation assumes that the future will repeat the past—the variables that have caused changes for the agencies in the past will continue to cause changes in the future. For example, the operations division of an

---

[2] See Barry Bozeman, *Public management and public policy analysis* (New York: St. Martin's, 1979), pp. 339–342.

agency might wish to forecast the agency's future energy use based on past usage, or a city manager may wish to forecast demand for sewage treatment. Although the principles of trend extrapolation are simple (they assume that causes of change in the future will be the same as in the past), the mathematics are fairly complex.

Correlational analysis, usually based on multiple regression technique, is increasing in popularity for forecasting. For example, a state welfare department may need to accurately forecast demand for its services for the next year. From past experience they know their caseloads are determined by the number of women in the population between the ages of eighteen and thirty-five (this correlates highly with Aid to Families with Dependent Children claims), the number of persons age sixty-five or older, and the number of unemployed. Using past data, the agency can regress welfare claims on the three variables. This will result in a regression equation. When the expected values for the number of women age eighteen to thirty-five, the number of persons age sixty-five and over, and the number of unemployed are entered, a predicted number of welfare cases results.

Despite the technical aura that permeates most forecasting, knowing what variables to forecast, the appropriate techniques to use, and the value of the completed forecast is an art. The manager should consider a forecast as an input into his or her decision-making process, and the forecast should be evaluated as one of many inputs. A series of future scenarios may be more valuable to the planner than one forecast with precise estimates.

SET PRIORITIES

Every program has more than one objective; programs have service objectives, efficiency objectives, responsiveness objectives, effectiveness objectives, and so on. For example, a recent CETA grant to a southwestern university called for a survey of Chicano service needs. The survey was to be conducted by the hard-core unemployed, retrained as survey-research interviewers. This program had training goals, future employment goals, effectiveness goals for both the training and the survey, cost goals, efficiency goals, and time goals.

Because an agency or a program will rarely be given sufficient resources to meet all its goals simultaneously, the various goals must be prioritized. For example, is it more important to have a valid survey of Chicano service needs or to train twelve survey researchers? Is it more important that the survey researchers be hard-core unemployed or that they be competent survey researchers? When objectives are ranked, trade-offs between the objectives can be considered. If an absolute ranking cannot be made, a general grouping of objectives into categories is also helpful.

DEVELOP ALTERNATIVES

The next stage of the planning process is to develop alternative methods to attain the agency's goals. In some program-planning circumstances, the number of alternatives is severely limited, while in others there are numerous strategic plan alternatives. Developing alternative programs is a difficult skill to teach. It requires creative individuals with substantive knowledge of the area who are not rigidly bound by past methods of doing the job. One way some organizations develop alternatives is through *brainstorming.* Brainstorming is a group process whereby individuals throw out ideas; other group members can embellish those ideas, but negative feedback is prohibited. The concept behind brainstorming is that quick, negative feedback makes people reluctant to express an unconventional idea, thus limiting the alternatives available to the group.

By developing a list of all possible alternatives before accepting or rejecting specific proposals, the agency gains the broadest possible perspective. Only after all possibilities are listed should a critical review occur.[3] The Delphi technique also can be used to generate program alternatives. Brainstorming allows the participants to explore alternatives in a free-swinging interaction. The Delphi approach, on the other hand, allows participants to explore alternatives in an unassembled fashion. Consequently, the Delphi method may be advantageous for agencies that are geographically dispersed.

EVALUATE THE ALTERNATIVES

Each of the proposed alternatives must then be evaluated in light of the expressed goals. The evaluation process can be a full-fledged, quantitative, systematic analysis of the alternatives or an intuitive, phenomenological evaluation or some combination of the two. Whether systematically or intuitively developed, alternatives are assessed in the context of such things as agency resources, secondary impacts on other agency programs, clientele acceptance, and political reality.

SELECT THE OPTIMAL ALTERNATIVE

In classical planning, the analysis continues until the optimal alternative is identified. In the real world, planners rarely seek one best alternative. Data and measurement may be too poor to permit sophisticated analysis. Goals

---

[3] See, for example, Andrew B. Van Gundy, *Techniques of structured problem solving* (New York: Van Nostrand Reinhold, 1981).

may be conflicting or ambiguously expressed. There may be insufficient time to analyze all the alternatives fully. Finally, political considerations may foreclose some options. In our normal incremental decision-making process, we therefore often select a satisfactory alternative, one that appears to be workable in the knowledge that other problems can be corrected the next time the program is reviewed.[4]

IMPLEMENT THE PLAN

Implementation is the phase in which program planning becomes operational. In order to implement, the agency must design an organizational structure; recruit personnel to administer the program; schedule the sequence of activities necessary for full implementation; procure needed funds and schedule their allocation; and provide mechanisms for feedback, control, and evaluation.

Successful implementation requires two elements: the unequivocal support of top management and a clear assignment of responsibilities for implementing elements of the plan. Without managerial commitment, the planning process becomes an exercise in futility, needlessly raising participant expectations and consuming time and energy resources that could be better spent on regular operations. Managers who engage in the motions of planning merely to placate dissenters or to boost morale may find that the process gets away from them as the participants' enthusiasm for change generates alternatives the manager is unwilling to implement. In such cases, it is better to do nothing than to stimulate new conflict and discontent.

A careful delineation of implementation responsibilities is critical because planning usually implies change. Change is usually resisted by those holding a vested interest in the status quo. Responsibility designation is also important because change calls for an extra effort by important organizational actors or a de-emphasis of other program components. Only by specifying who will do what and who is answerable to whom for plan completion can a plan involving change hope to succeed.[5] (See Chapter 6 for a complete amplification.)

Earlier we alluded to the necessity of evaluating plans in the context of an agency's environment, particularly the political climate. Politically, an agency may not be able to implement its ideal plan, but having plans ready when the political climate becomes more favorable allows an agency to move at the opportune moment. In the post-Vietnam era, for example, political decision makers were unreceptive to new weapons systems, military pay increases, and

---

[4] Herbert A. Simon, *Administrative behavior* (New York: Free Press, 1966).

[5] Fred Beckhart and R. T. Harris, *Organization transition: Managing complex change* (Reading, Mass.: Addison-Wesley, 1977).

the like. However, the climate changed quickly, partly because of Soviet adventurism in Afghanistan and the taking of American hostages in Iran. Suddenly, military expenditures were perceived as necessary and desirable. The speed with which Defense Department officials were able to present decision makers with a full range of detailed defense alternatives can be credited to the planning emphasis in the Department of Defense, an example that should not be lost on other agencies.

EVALUATE THE PLAN

Planning outputs, especially the programs that result, should be subjected to the evaluation process. The result of the evaluation provides feedback to management on program performance and allows completion of the final step, revision of the plan. The techniques of evaluation are discussed in Chapter 5.

REVISE THE PLAN

Planning is a continuous process. As the evaluation results become available, they can be used to restructure goals and objectives. Where new needs arise or second-order consequences appear, the planning process should incorporate them. When managers know they will get a second shot at a problem, they will be more willing to attempt a creative solution. Planning therefore should be thought of as a continuous process.

SOME CAVEATS ABOUT PLANNING

The planning process contains certain pitfalls of which the manager must be aware in order to minimize their impacts.

DIFFICULTY IN ATTAINING CONSENSUS ON OBJECTIVES

If two members of Congress were asked why they support a negative income tax (NIT), one member might respond that the NIT will eliminate the restrictions welfare currently places on individuals' lives (choice about marriage, family desertion, how to spend money) and therefore would be a major liberal reform. Another member might respond that the NIT will turn welfare into a transfer payment program and therefore eliminate thousands of government employees; in this member's eyes, it would be a major conservative reform.

This example illustrates a common problem. People can often agree on programs yet disagree about goals. Seeking consensus on goals is seeking consensus on highly committed values. Conflict is almost always present because each portion of the organization will see the problems of the organization differently. As Rufus Miles (a former BOB official) wryly stated, "Where one stands depends on where one sits."

OVERSIMPLIFICATION OF PROBLEMS

Because top-level planners cannot consider all the nuances of an agency's programs, they may ignore many of the details and gloss over the complexities of implementation. Simplification enables them to plan, but it also weakens the planning process. One way to minimize the possibility of oversimplification is to expand the planning group to include a broad spectrum of organization personnel.

TIME CONSTRAINTS

Planning suffers from both the lack of time to devote to it (the alligator problem) and the lack of time to conduct analyses of alternative programs. As a result, planning may be avoided or done with such haste that little analysis is possible. Effective planning requires a commitment of sufficient resources to do it right.

INFORMATION CONSTRAINTS

Because good information, especially in regard to forecasts, is often not available to the planner, whatever information is available becomes more important. When decisions must be made, timely information takes precedence over quality information. One reason the General Accounting Office (GAO) has its degree of influence in the congressional policy process is that its inputs are always timely. Of course, agencies frequently take issue with the quality of the GAO analyses.

COGNITIVE NEARSIGHTEDNESS

Decision makers show a preference for quick short-term results over long-term solutions that may be superior. This happens in part because agencies tend to be responding to crises and in part because of the expected short tenure of many decision makers (next year it's someone else's problem). An example of this shortsightedness is the Old-Age, Survivors, and Dependents Insurance program changes of 1977. Faced with a revenue shortfall, decision makers opted to increase social security taxes and solve the immediate problem, rather than redesign the system to avoid future revenue problems.

POLITICAL CONSTRAINTS

Agency decisions must conform to political realities even when politics conflict with rational planning decisions. Clientele, Congress, and elected executives are all capable of limiting the options available to the agency. The Army Corps of Engineers, for example, has often been charged with altering its cost-benefit analyses to fit political realities. On the other hand, Congress may require a specific program and even instruct the agency on how the program is to be administered. The Environmental Protection Agency's plans for air quality, for example, are restrained both by congressional action and by technological feasibility.

MEANS VERSUS ENDS

Planning is generally concerned with values—with the ends of government programs. In the public sector, not only must the ends be laudable but the means must be acceptable. An agency must administer a program that treats its clientele humanely, keeps adequate financial records, and considers the expertise of its employees. When ends are established by law, variations in means often become the important variable. In general, planning de-emphasizes means but it need not do so if the plan contains a process-evaluation component.

VALUES OF PLANNERS

Planners are no different from other decision makers. Despite the norms of objectivity in the planning profession, the planners' values may well influence the outcome of a plan. Michael Vasu's study of the planning profession revealed that planners generally held values different from those of the general population. Where managers fail to specify goals and objectives clearly, planners may substitute their own.[6]

In some instances, however, value-free planning is not desirable. *Advocacy planning,* as it is called, has gained recent popularity among those who wish to ensure that varied interests are represented in the planning process. An advocacy planner, for example, might be called on to design a military retirement system

---

[6] Michael L. Vasu, *Politics and planning* (Chapel Hill: University of North Carolina Press, 1979).

that maximizes benefits for enlisted personnel or to critique all proposed alternatives from the perspective of enlisted personnel.

The foregoing caveats are presented to divert the novice planner from some of the pitfalls common to the planning process. They are not intended to denigrate the value of planning, which is an activity we believe to be intrinsically worthwhile. Now that the reader has a grounding in the general concepts of planning, we turn to a more detailed presentation of the various types of planning in which government agencies engage. Although all types of planning are discussed in a hierarchical order, the emphasis is on program planning, which is the level at which most planning occurs.

THE PLANNING HIERARCHY

In general, plans form a hierarchy. Those plans at the top tend to be general in scope, covering a great many aspects of organizational activities. In some instances macrolevel planning cuts across agency lines. Because lower-order, operational plans have greater detail and are more useful to the line manager, operational planning is emphasized here.

NATIONAL PLANNING

At the top of the hierarchy are plans for the entire nation. Because five-year plans and official national plans raise the specter of socialism, American planners do not refer to their plans as national. For most of its history, the U.S. government has not planned or has at least consciously resisted systematic program planning. The most notable exception to the no-planning dogma was the attempt by Lyndon Johnson's administration to use a budget system known as Program-Planning and Budgeting Systems (PPBS) as a vehicle for planning federal programs. PPBS failed for a number of reasons, including bureaucratic and political resistance, the complexity of the process, and the difficulty of applying the techniques of cost-benefit analysis to nontangible programs such as welfare and human services.[7]

Opponents of PPBS argued that the system went against the grain of American political values, which hold that the proper method for making public policy is the give-and-take of public debate in which policy antagonists work out compromises involving incremental changes in current policies rather than

---

[7] For a fuller discussion, see Stanley B. Botner, "Four Years of PPBS," *Public Administration Review,* July-August 1970, pp. 423–431.

wholesale changes. Planning opponents argue that projecting long-range plans and committing resources for five years or longer has the effect of denying future policy makers the right to debate the issues affecting their era.[8]

Public agencies, particularly those requiring the development of new technologies, must plan. Weapons systems are generally obsolete before leaving the drawing board. Similarly, the programs of the National Aeronautics and Space Administration (NASA) require planning years in advance. Nontechnology-based agencies also must plan. For example, the Social Security Administration must project expenditures and revenues, and state and federal transportation departments must plan for the development of new highway systems as well as for the repair and replacement of existing roadways.

Even agencies for which planning is essential may be frustrated by the changing winds of politics. Two excellent cases in point are the on-again, off-again plans for the deployment of the B-1 bomber and the MX missile systems. One might erroneously conclude from these examples that it is impossible to plan in government. To the contrary, the vagaries of politics make it necessary for an agency to generate a variety of contingency plans. The absence of a range of plans can result in outright loss of funding or the diverting of a portion of an agency's resources to competing agencies that possess a variety of contingency plans to present to decision makers.

AGENCY PLANNING

For most public servants, national plans are fairly remote. The first level of planning that affects most administrators directly is the agency-level plan. Agency plans can be either strategic or operational.

Strategic Plans. A strategic plan is one that maps out methods of obtaining broad agency goals. These goals may concern program jurisdiction, interest group relationships, congressional allies, and so on. A strategic plan considers more than how the agency can best perform the tasks assigned to it. Strategic plans concern the agency's domain. What functions the agency should perform, who the agency should serve, and how the agency can maximize its resources are all examples of domain-relevant questions. A good strategic plan is also a prerequisite to optimal operational planning.

Operational Plans. Operational plans focus on program activities that can be managed within the agency. For example, the Old-Age, Survivors, and Depen-

---

[8] Aaron Wildavsky, *The politics of the budgetary process,* 2nd ed. (Boston: Little, Brown, 1974).

dents Insurance program generates actuarial plans involving different birthrates and deathrates and various demographic trends. This plan forecasts problems that will affect the system. Alternative solutions can then be offered to Congress before problems arise. The remainder of this chapter focuses on how to carry out operational plans.

PROGRAM PLANNING

One type of operational planning is called program planning. In program planning, the manager is given a set of goals and possibly some broad program parameters. The manager then must design a system for successful program implementation. Program planning emphasizes organizing, scheduling, budgeting, and controlling. Several techniques exist to improve program planning. Whereas strategic planning is almost always considered an art, many people believe that program planning is a skill.

UNIT PLANNING

At the lowest level of planning are unit plans. These may be plans for one aspect of a program or for a single unit of the organization. Although the distinction between program plans and unit plans is somewhat artificial, unit plans tend to be short-range and limited. An agency's affirmative action office, for example, may plan to hire five minority members and seven women in the next year. This goal, plus the methods the office uses, would be a unit plan.

USING PERT/CPM FOR PROGRAM PLANNING

One of the more formalized operational planning strategies is known as Program Evaluation and Review Technique (PERT) or the Critical Path Method (CPM). PERT/CPM came into use as a system of project management for projects with definite start-up and termination dates. PERT/CPM gained widespread notoriety as the system used by NASA to manage the manned space-flight programs as well as other NASA projects, but it has application for social-program planning as well as engineering projects. The linearity of PERT/CPM, along with its clear-cut timetable, makes it appropriate for making new social programs operational and for carrying out the thousand-and-one special projects that must be completed in addition to the regular mission of the agency.

PERT is one of several forms of network analysis that are used in program planning. A variety of embellishments have been made on the basic version of PERT, but the fundamental concepts remain the same.

THE PERT CHART

A PERT chart consists of a series of events (designated by circles) connected by the activities necessary to complete the events (designated by arrows). Above each arrow, time estimates (for PERT-time) or budget estimates (for PERT-cost) are written. (Although our discussion of PERT here focuses on PERT with time, PERT can also be used to schedule and control costs. The only difference is that the activities have expected costs rather than expected times.) Events represent specific activities that have begun or been completed. Event circles are not to be used for processes; all processes must be designated by activity arrows. In the following example, two events, "Begin Cost-Benefit Draft" and "Complete Cost-Benefit Draft," are connected with an activity arrow that signifies the writing of the cost-benefit draft. The numbers above the arrow are time estimates

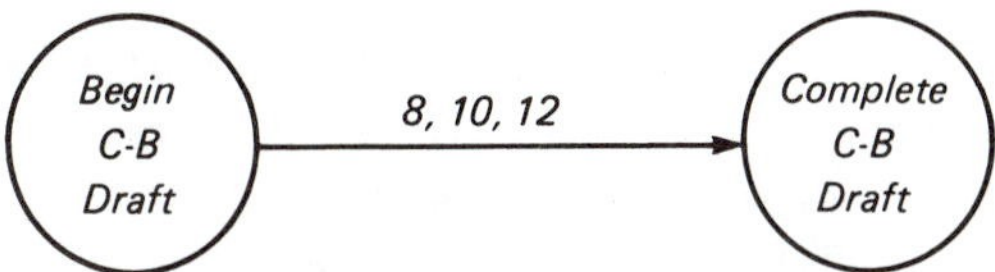

to complete the draft (in this case in days). The first estimate, eight days, is the optimistic estimate, the shortest possible time to complete the activity. The third estimate, twelve days, is the pessimistic estimate, the longest time to complete the activity. The middle figure, ten days, is the most likely time it will take to complete the activity. These three estimates are usually combined into one estimate called "Expected Time" with the following formula:

$$\text{Expected Time} = \frac{\text{Optimistic} + 4\ (\text{Most Likely}) + \text{Pessimistic}}{6}$$

In this case,

$$\frac{8 + 4 \times 10 + 12}{6} = 10 \text{ Days}$$

Some PERT charts retain the three estimates; others just list the expected times. In this chapter, to simplify the drawings, we will use only the single estimate.

PERT charts operate according to serial logic. All events logically prior to an activity must be completed before that activity can start. In the following chart, activity *c* cannot be started until event *X* is completed. Since activity *a* cannot be started before the completion of event *V*, both *V* and *X* must be completed before activity *c* can begin. Note that event *W* has no affect on activity *c*. Two or more activities may lead to a single event, as *c* and *d* lead to event *Y*. Note

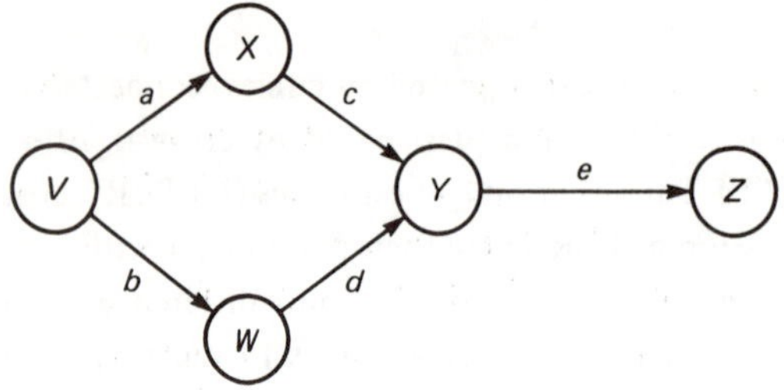

that the activities flow in one direction. Feedback loops are not permitted because they are inconsistent with the serial logic. For example, an activity arrow could not run from *Z* to *W* because that would imply that *Z* must be completed before *W* and, at the same time, that *W* must be completed before *Z*.

Figure 2-1 illustrates several other PERT concepts. This PERT chart represents the preparation of a teaching manual to be used in a program planning and evaluation course. All the estimated times have been calculated.

The first step in developing a PERT chart is to calculate the total expected time to complete all tasks. Each "put-in" (activity) of the chart will have a different expected time, so all put-ins must be calculated. For the chart in Figure 2-1, the top put-in runs through events *a, b, c, d, g, m, o, r,* and *s.* Summing the time values for the activities reveals an expected time of 19.2 days. The remainder of the expected times are:

Path	*Route*	*Expected Time*
2	*abcdhmors*	25.9 days
3	*abceinprs*	24.4 days
4	*abcfkqrs*	17.5 days
5	*abclqrs*	8.3 days

THE CRITICAL PATH

Using the times for each put-in, the analyst selects the path taking the most time. This is called the critical path. In Figure 2-1 the critical path runs through events *a, b, c, d, h, m, o, r,* and *s.* Often the critical path is shown by a double line or a colored line to make it more visible. Any time delays on the critical path will delay the entire project. The critical path tells us that the entire project will take 25.9 days. At this point the manager needs to know how much time is available. If there are 26 or more working days between the contract award and the final delivery date, the PERT time estimates are acceptable. If fewer days are available, say 22 (less than the PERT chart requires), the program must be redesigned to finish within 22 days.

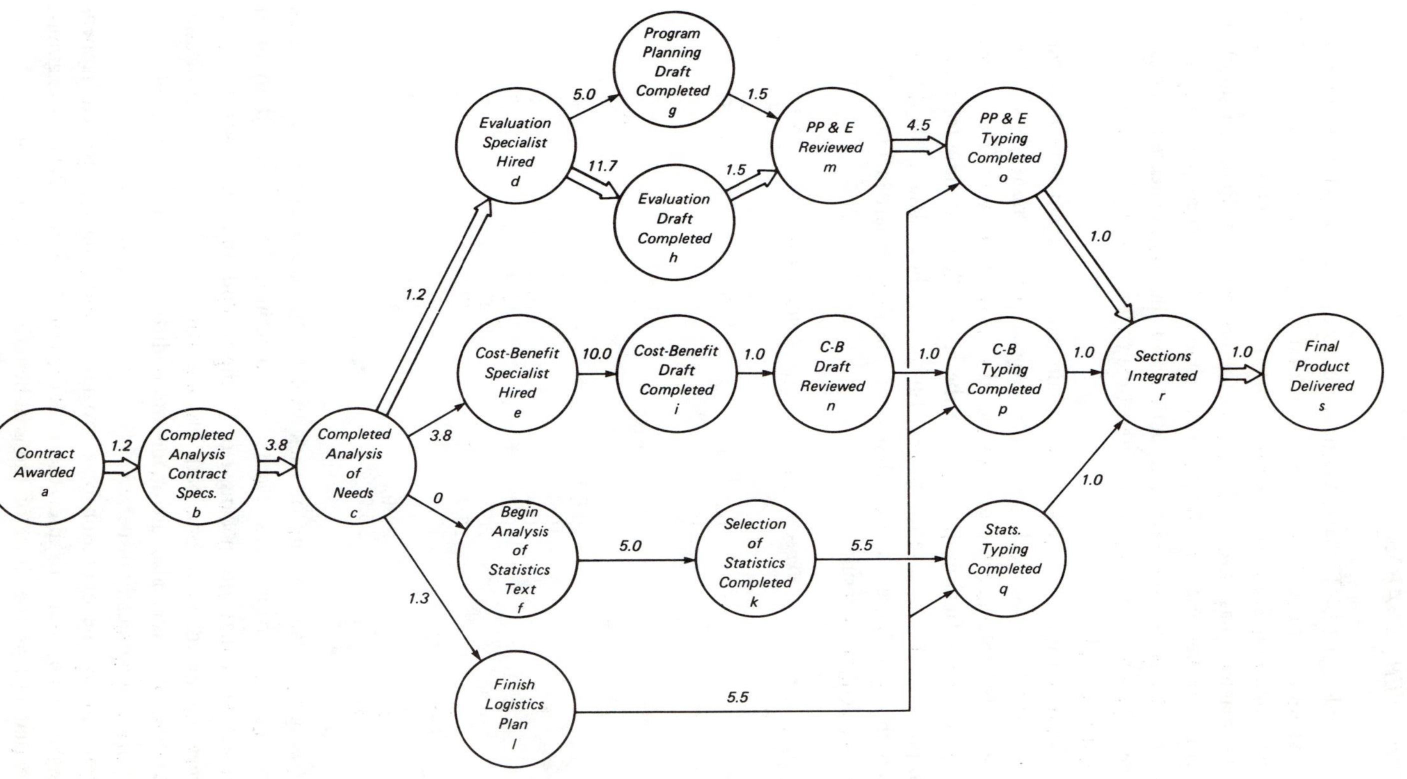

FIGURE 2-1
PERT chart for design of program planning and evaluation manual

ALTERING PERT CHARTS

When we redesign PERT charts we must focus on the critical path since it requires the longest time. In this case, a second person could be hired to help write the evaluation draft. If, as a result, the time is shortened from 11.7 to 6.0 days, the manager has saved 5.7 days on the critical path. Redesign should be done with care because adjustments may change the critical path. In our example, the change is from path 2 to path 3 and the total expected time is still 24.4 days. To reduce the total expected time to less than 22 days, other adjustments must be made.

All events not on the critical path contain slack. An event with slack can be delayed without setting back the entire project. To determine slack time for each event, first calculate the total expected time for each event. Then, working backward from the final event, calculate the latest allowable time for a task to be completed and still have the project finished without delay. The difference between the two figures is slack time. For the PERT chart in Figure 2-1, the following calculations result:

Event	*Expected Time*	*Latest Time*	*Slack*
a	0	0	0
b	1.2	1.2	0
c	5.0	5.0	0
d	6.2	6.2	0
e	8.8	10.3	1.5
f	5.0	13.4	8.4
g	11.2	17.9	6.7
h	17.9	17.9	0
i	18.8	20.3	1.5
k	10.0	18.4	8.4
l	6.3	18.4	12.1
m	19.4	19.4	0
n	20.6	22.1	1.5
o	23.9	23.9	0
p	22.4	23.9	1.5
q	15.5	23.9	8.4
r	24.9	24.9	0
s	25.9	25.9	0

Knowing the exact amount of slack available for each event can be beneficial because resources from slack events can sometimes be transferred to other project activities that are in danger of falling behind. In many cases, however, resources contained along one path are specialized and therefore not easily transferable. Transfers also are made more difficult when components of a project are geographically dispersed.

Often, during the operation of a program, something unexpected happens that makes it necessary to alter the PERT chart. In Figure 2-1, for example, assume that just as the draft of the cost-benefit section is complete, the cost-benefit specialist dies of Legionnaires' disease. Since a cost-benefit specialist is still needed, this will delay subsequent events on this path. Unexpected events are affectionately known as PERTurbations. When a PERTurbation occurs, actions are taken to remedy the problem and adjust the PERT chart. One adjustment could be as follows:

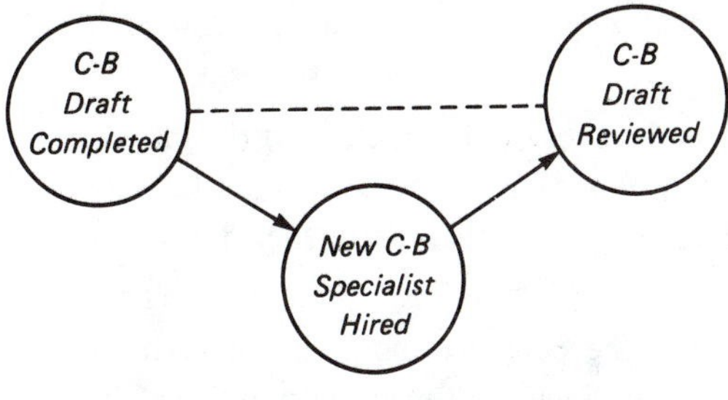

MONITORING A PROGRAM WITH THE PERT CHART

The PERT chart is often used to monitor the status of a program and flag any potential problems. Many sophisticated computer programs exist to monitor PERT. For a small project, the easiest way to use PERT is first to translate the PERT chart into a schedule that lists the days that events are to begin and are to be completed. As events are completed, the event circles in the PERT chart are colored in. This permits the manager to see the status of any event at a glance. More sophisticated charts are drawn so that the length of each activity line fits a time scale. In this manner all the events to be completed on the same day fall along a vertical line. With such charts, a movable string or line is placed on the chart at the current date. This permits the manager to see at a glance if all paths are on schedule and to locate any problems quickly.

SETTING UP A PERT CHART

Ross Clayton of the University of Southern California has developed a step-by-step process for setting up a PERT chart.[9] Although his is not the only method, it is a concise, well-structured procedure.

[9] Ross Clayton, "Techniques of Network Analysis for Managers," in *Managing public systems: Analytic techniques for public administration,* eds. Michael J. White, Ross Clayton, Robert Myrtle, Gilbert Siegel, and Aaron Rose (North Scituate, Mass.: Duxbury Press, 1980).

1. The project manager clearly specifies the project's final objective. What will the last event in the PERT chart be? In setting up other portions of the chart, it often helps to think backward from the final event.
2. The project manager should establish a set of working assumptions: How long will the project last? What will the budget be? Specifying the working assumptions means that all participants will be working on the same problem.
3. The project manager develops the subsystem structure of the program. What will the major paths be? For example, what are the key components of the program? These paths can then be further divided into subpaths.
4. The project manager assembles the people responsible for each path into a program-management team. Each path manager may have his or her own mini program team.
5. The program management team decides how detailed the PERT chart should be, what cost or time guidelines apply, and what the common planning format will be.
6. The project manager, in conjunction with the path manager, develops the PERT diagram for each path.
7. The project team combines the PERT paths into an overall program PERT chart.
8. The project manager circulates the PERT chart to the management team and revises the events and activities to form the completed PERT chart.

OTHER ISSUES IN PLANNING: A MANAGER'S CHECKLIST

There are certain indispensable components of effective planning that are frequently overlooked by managers. Some of them were explicitly or implicitly discussed in preceding sections. Their emphasis here exemplifies their importance to successful planning.

PROGRAM DEFINITION

The people responsible for a program, and those above them in the chain of command, need to define the essentials of the program. The first topic of discussion should be program goals and objectives. What is the program seeking to achieve? How will the manager know if the program works? For a program without a clearly specified output, this step in the process would define the functions that the program should perform for the organization.

As part of the program definition, program authority and priority need to be specified: Under what authority is this program being conducted? Specific attention should be given to applicable legislation or to administrative decisions. At this point, the organizational leadership should assign an overall organizational priority to the program: Is this a program that is vital to the organization or is it being conducted just to please some outside interest? Even if the program priority is not specified in writing, it should be communicated to the program manager. This allows the program manager to assess the relative value of the program to the organization and also indicates how much organizational support is possible.

The third aspect of program definition is the availability of resources. If possible, resource figures should be specified for the life of the program rather than just for one year. This permits more comprehensive cost planning. Other resources, such as personnel position allocations and physical facility requirements, should also be included. These constraints may be drawn from previously obtained figures or they may be ballpark estimates to be fleshed out by the program manager.

STRUCTURE DECISIONS

After program definition is completed, a series of structural questions relate the program to the organization and its overall purposes. The first structural question concerns reporting relationships: Who will supervise the program (that is, to whom will the program manager report)? This question must be answered not only for the substance of the program but also for budgets, personnel, auditing, and so on. This structural aspect places the program in the organization's chain of command.

The second structural issue is the organization of the program itself. Several options are available. The program can be organized in a strict scalar chain of command with firm reporting guidelines, or it can be loosely structured in matrix management form (see Chapter 6). As an alternative, the structural decisions may be left to the program manager. The number of structural options is virtually unlimited. Given the conflicting nature of the findings on structure and performance, managers should select a structure that fits the program's environment, the people who will work in the program, and the manager's own preferences.

PERSONNEL DECISIONS

The next set of program-planning decisions concerns program personnel. The most important of these decisions is the definition of the project manager's role (or job description): What decisions can the program manager make? What decisions must be cleared at a higher level in the chain of command? What decisions can be delegated? Does the manager have full control over all aspects of the program or is the manager simply an administrative coordinator? The supervisors of the program manager may decide that flexibility is essential and may operate under any one of several versions of management by objectives (MBO). In other circumstances, where process is important, the duties and actions of the program manager may be spelled out in detail.

Included in the program manager's job description should be the manager's reporting relationships. These may or may not be different from the reporting relationships of the program because the manager may not have responsibility for all aspects of the project (for example, on some technical projects the manager is responsible for administrative matters but not for technical performance).

The final aspect of the program manager's job description is the qualifications necessary to do the job. The description can therefore be used both to recruit the manager and as an official job description. At this point, staffing decisions about the program manager should be made if the manager has not already been selected.

DEVELOP A PERT

Program planning does not always involve the use of PERT, which is most appropriate for special projects of a limited duration. When PERT is appropriate, however, the planning group would develop a PERT time chart and a PERT cost chart at this point.

EVALUATION DECISIONS

The final stage of program planning is evaluation of the plan. The evaluation contains three parts—feedback, audit, and program evaluation. Feedback specifies the reports that will be required from the line managers to the program director. This portion of the evaluation is merely the establishment of a mini management information system. The amount and form of the feedback depends on the program manager's personal preferences and the audit and program evaluation needs.

The audit requirements often are set by law. If they are not, the program manager needs to design a postaudit strategy to ensure that the program is in compliance with pertinent laws.

The program evaluation plan is a tentative design for evaluating the success or failure of the program. This plan should include the research design, specification of measurable indicators of program goals, provisions for data collection, and the assignment of responsibility for conducting the evaluation.

The appropriate time to begin specifying the parameters of the program evaluation is during the program-definition phase, prior to implementation. By specifying the evaluation format and criteria in advance, program officials allow for the collection of baseline data or current program performance against which the new or adapted program can be compared. Thus we come full circle. At the outset of the chapter we emphasized the importance of program planning to program evaluation. We conclude by pointing out that carefully defined evaluations are part and parcel of artful program planning.

Case Studies

WETLANDS, WATERFOWL, AND WAIVERING RESOURCES

Each year twenty to thirty million Canadian ducks migrate to the fertile San Joaquin Valley to pass the winter in the relatively mild climate. At one time, much of the valley was wetlands, where the one to two feet of water depth provided a perfect habitat for the ducks. Modern agricultural growth and other incursions of humans, however, have drastically reduced the amount of available wetlands.

To offset the loss of habitat, the Fish and Wildlife Service of the federal government artificially floods acreage in the northern and south-central regions of the valley. The larger preserve is located at the northern end of the valley near Sacramento. The other preserve is the Kern National Wildlife Reserve, located near Bakersfield, California.

The principal human users of the preserves are hunters. (A growing number of bird-watchers have also begun visiting the preserves.) Congress has provided that a portion of federal hunting fees be allocated to the operation of the Fish and Wildlife Service. Understandably, the service is aware that hunting groups are among their principal clientele.

The Sacramento preserve can accommodate up to 250 hunters per day. The Kern preserve can accommodate 80 hunters. In addition, a number of private hunting clubs have been built adjacent to the preserves. Although it can accommodate far fewer ducks and hunters, the Kern preserve serves a population larger than that served by the Sacramento preserve. The Kern preserve is approximately 100 miles from the Los Angeles basin. The next nearest accessible hunting area is more than 200 miles away.

Recently, increased energy costs have escalated the cost of flooding the acreage on the Kern preserve. In 1980, flooding cost $17 per acre foot. By 1982 the cost had risen to $28 per acre foot. By contrast, 1982 pumping costs at the relatively wetter Sacramento preserve were $2.50 per acre foot.

The Fish and Wildlife Service must decide whether to continue operating the more expensive Kern preserve. It is having difficulty absorbing the increasing pumping cost during a period of government austerity and, if the preserve were closed, pumping and operating costs could be reallocated to the more cost-effective northern operations. In making its decision, the Fish and Wildlife Service must consider the following: The preserve services between twenty thousand and three hundred thousand ducks out of a total annual population that varies between twenty and thirty million; discontinuing the Kern preserve would reduce the amount of available wetlands, although it is assumed that the ducks would find other places to winter; not pumping the water into the wetlands would increase the supply available for agricultural uses in a region that is relatively arid; the reaction of the hunting population may well be unfavorable.

Utilize the following decision model and

1. Determine the goals of the organization.
2. Develop a list of alternatives to resolve the problems of the agency.
3. Define the changes in the organization's environment.
4. Define the relevant actors in the agency's environment.
5. Define the components of the agency's mission and delineate real or potential problems.
6. Reassess the alternatives in light of the environment.
7. Choose an alternative on the basis of the reassessment.

A PERT APPLICATION: THE CASE OF DUSBOW, OKLAHOMA

Dusbow, Oklahoma, is a city of 150 thousand located in the western portion of the state. Dusbow enjoys the distinction of having been the western-most outpost of the U.S. Cavalry for a number of years. What was once a small military post has since grown to be one of the U.S. Army's principal training centers. Consequently, Dusbow has an additional population of some eighty thousand men under the age of twenty-five. The impact of this increased population of young men without roots in Dusbow has not been lost on the police chief. Such a concentration of young adults makes crime a problem in Dusbow disproportionate to the city's size.

The city's chief of police came up through the ranks of the department in a career that spanned twenty-four years. During his rise, he acquired a bachelor's degree from the local university, and he is currently working on a master's degree in criminal justice management from the state's largest university. In his graduate study, the chief learned about two new concepts in police work: the career criminal program and the victim/witness service center.

Career criminals are repeat offenders whose principal source of income is stealing from others. Career criminals are also wise to the workings of the criminal justice system. In fact, career criminals are so wise that they are rarely arrested; they always have a set of ready answers to offer suspecting officers. If arrested, career criminals never give their right names or addresses. Thus, record checks are difficult, especially if the offense for which an arrest is made is minor.

If caught in the act of a serious crime, the career criminal frequently will make bail and leave the area. If bail cannot be arranged, the career criminal will insist on plea bargaining with prosecutors for a reduced charge and sentence. For their part, overworked prosecutors frequently welcome the opportunity to have criminals plead guilty to lesser charges, thereby saving everyone the effort and expense of a jury trial.

Because of their sophisticated knowledge of the system, career criminals often avoid detection, detention, and lengthy incarceration. A cadre of career criminals in a city the size of Dusbow might number no more than fifty but account for 70 percent or more of property crimes. This is not to suggest that career criminals do not also commit acts of violence; to the contrary, career criminals are frequently responsible for an entire range of violent acts that usually go unpunished.

A career criminal program is one way law enforcement officials have developed to deal with the problem. Such a program involves training officers to recognize and arrest career criminals, and the special handling of career criminals in the justice system to guarantee their right to a speedy trial. In addition, prosecutors focus on ensuring that the due process rights of career criminals are not violated and on seeking the maximum sentence. The goal of the career criminal movement in law enforcement is to put career criminals in prison for as long as possible.

Persuading citizens to come forward and testify against criminals is a second problem facing police administrators and prosecutors. Witnesses may refuse to get involved for a number of reasons. These may include a distrust of authority, court schedules that interfere with witnesses' work schedules, insensitive officials who do not distinguish between criminals and witnesses during their investigations, and the need for childcare and transportation during the trial process.

The principal source of problems for victims and witnesses is the attitudes of investigating officers. These officers all too frequently do not differentiate between victims/witnesses and criminals. The officers, therefore, are tough and aggressive toward all present at the crime scene. The problem is compounded when officers harbor prejudices against members of minority groups. When members of a minority group are victims or witnesses of crimes by other members of the same group, the former may be reluctant to assist surly white police officers who treat them as if they were the criminals.

Because victims and witnesses of crime are frequently members of minority groups and/or poor, they may distrust police and legal institutions. Persuading them that the peacekeeping mechanisms of the society can and will act in their interest is a task of the first order of difficulty. Breaking down these fears is a necessary first step in improving authority-community relations sufficiently to enlist victim/witness assistance in prosecuting criminals.

The logistical problems of participating in the criminal justice process are not confined to the poor. Middle-class citizens also find the repeated delays in the process a source of frustration. All too frequently, the defense or prosecution will postpone the trial date several times. Each delay makes it necessary for victims or witnesses to take another day off from work, find transportation to the courthouse, and/or arrange for child care. Sophisticated defense attorneys may even use delaying tactics until they determine that all the witnesses are not present, at which time the attorney's will demand that the trial process begin immediately in accord with the client's right to a speedy and just trial. If prosecution witnesses are not present, the judge is likely to dismiss the case.

Several federally funded pilot programs for improving relations with and service to victims and witnesses of serious crimes have been tested. The concept is simple. The local prosecutor and police department develop a jointly operated victim/witness service center. The purpose of the center is to provide free transportation for trial participants to and from court and child care as necessary. The center also runs training seminars to sensitize police to the impact their attitudes have on citizen support in criminal investigations. Finally, the center operates coordinating services for scheduling court appearances of all concerned parties. The coordination unit secures commitments from victims, witnesses, prosecutors, and defense attorneys that they will appear at the same place and time to conduct their mutual business. Defense attorneys who agree to appear and then withdraw are brought to the attention of the trial judge. If a delay is necessary, interested parties can be notified a day or two in advance and avoid the frustration of appearing and being sent home.

Both the victim/witness service center and the career criminal program have been undertaken as pilot programs funded by the federal government in recent years. Both programs have yielded significant improvements in the conviction rates of criminals and, in some cases, substantial drops in the crime statistics for the cities in which the programs have been undertaken.

Unfortunately for Dusbow, there are no more federal funds for either program. If Dusbow wants to engage in either or both, the funds will have to be found locally.

The Dusbow chief of police is convinced that both programs are worth undertaking in his city. The chief has noticed that pilot programs for victim/witness service centers and career criminal programs have rarely been tried together, but he believes that a combination program would be easier to sell and would result in dramatic improvements in the law enforcement and criminal justice systems of Dusbow.

The city manager of Dusbow is a progressive person who has managed to sustain good relations with the city council for eight years, despite efforts of police and fire unions to win in the council chamber what they could not secure at the bargaining table. The manager is readily convinced by the police chief that a combination program would be worthwhile if funding can be secured. Both the manager and the chief visit the district attorney and convince the prosecutor that the program would be beneficial to the community and to his conviction rate and subsequent chances for reelection.

The agreement to develop a combined program is reached three months before the new budget cycle for the city begins. The prosecutor, the police chief, and the manager agree to form a program-planning task force to (1) develop a fundable victim/witness center and career criminal program before the start of a budget hearing in ninety days and (2) work together for the enactment of the plan. All parties agree that the plan can be funded and fully operational six months from the date of council authorization.

Develop a PERT of the functional components of a combination victim/witness and career criminal program to be brought on line in nine months. The plan should include but not be limited to:

- Establishment of the victim/witness service center
- Training programs for police and prosecutors
- Establishment of judicial liaisons
- A public relations campaign to enlist citizen support
- Establishment of special police/prosecutor units for prosecuting career criminals

FOR FURTHER READING

Ascher, W. *Forecasting: An appraisal for policy makers.* Baltimore: Johns Hopkins University Press, 1978.

Bazaraa, M. S., and Jarvis, J. J. *Linear programming and network flows.* New York: Wiley, 1970.

Beckhart, F. *Organization development and strategies and models.* Reading, Mass.: Addison-Wesley, 1969.

Beckhart, F., and Harris, R. T. *Organization transition: Managing complex change.* Reading, Mass.: Addison-Wesley, 1977.

Clayton, R. "Techniques of Network Analysis for Managers." In *Managing public systems: Analytic techniques for public administration,* edited by M. J. White, Ross Clayton, Robert Myrtle, Gilbert Siegel, and Aaron Rose, pp. 86–107. North Scituate, Mass.: Duxbury Press, 1980.

French, W. L., and Bell, Cecil H., Jr. *Organization development: Behavioral science interventions for organization improvement.* 2nd ed. Englewood Cliffs, N. J.: Prentice-Hall, 1978.

Morris, L. N. *Critical path construction and analysis.* New York: Pergamon Press, 1967.

Ostrofsky, B. *Design, planning, and development methodology.* Englewood Cliffs, N.J.: Prentice-Hall, 1977.

Chapter 3

COST-BENEFIT ANALYSIS

Cost-benefit analysis is an analytical tool used to study the impact a project will have. It can facilitate decision making regarding adoption, implementation, or continuation of a project. Cost-benefit analysis has been widely used to compare and choose among alternative proposals for meeting a specific goal, but it is also used to evaluate the results of past programs. The general approach is to identify and quantify both the negative impacts (the costs) and the positive impacts (the benefits) of a proposed project and then to subtract one from the other to determine the net benefit—thus the name "cost-benefit analysis." Of course, it is not that easy, or project evaluation would be a breeze. But the appeal of cost-benefit analysis, and one of the major reasons for its unintentional as well as intentional misuse, lies in the simplicity of the concept of comparing gains and losses. After all, most of us are taught to do this type of trade-off analysis when solving personal and professional problems. The difficulty comes in the application of cost-benefit analysis to large-scale government projects whose costs and benefits are seldom clear to the evaluators or the interested public.

Because cost-benefit analysis varies according to the scope, size, substance, and purpose of the project to be evaluated, a formula approach cannot cover all instances. However, those wishing to use or understand this tool must become familiar with (1) the basic rationale for using cost-benefit analysis, (2) the major assumptions required for proceeding with a cost-benefit analysis, (3) the general criteria used for making a final decision based on cost-benefit analysis, (4) the economic theory and economic concepts from which cost-benefit analysis has evolved and which dominate it, and (5) some of the major problems that arise when one attempts to apply these concepts to social policy decisions.

Cost-benefit analysis is the generic name for the analytical model used in this chapter. A variety of names are used to describe cost-benefit analysis or project evaluations that use cost-benefit techniques. Most of these are simply subtypes of cost-benefit analysis. For example, risk-benefit analysis has recently received a great deal of attention because it has been used in evaluating projects such as nuclear power plants, the siting of new chemical plants, and the need for air bags as safety devices in automobiles. These studies essentially have been cost-benefit analyses that share the unique problem of having to value human life as a primary component of the loss-gain formula. In addition, they must deal with a high degree of uncertainty and/or risk, using probabilities as substitutes for facts or experience. Cost-risk-benefit analysis is another name for this type of evaluation. Cost-effectiveness or cost-effective analysis is another subtype of cost-benefit analysis in which the costs (inputs) are measured against outputs. The resulting ratio, indicating the "effectiveness" of the project in using available resources to achieve a desired output, is then compared with the effectiveness of the other alternatives utilizing the same inputs. The ratio can also be used when benefits of alternative projects are judged to be the same. In this instance, the costs incurred by each should be compared across projects. Cost-effectiveness is often used as a component of a broader cost-benefit analysis. The term that is perhaps most frequently used in cost-benefit analysis is *benefit/cost ratio,* which is simply the total benefits divided by the total costs. Based on predetermined criteria, the benefit/cost ratio makes it possible to choose the alternative with the highest positive ratio of benefits to cost.

The opening discussion of why cost-benefit analysis is so widely utilized is followed by the assumptions on which cost-benefit analysis is based. The actual techniques of cost-benefit analysis are then presented. This discussion explains how projects should be identified, how to develop a list of potential impacts of a proposed project, and how the costs of various project components are estimated. Pricing techniques are presented, followed by the use of discount rates in cost-benefit analysis. After a presentation of the criteria used in comparing costs and benefits, the chapter concludes with an overview of the role of the cost-benefit specialist, cost-benefit analysis, and a discussion of the pros and cons of using cost-benefit analysis.

WHY USE COST-BENEFIT ANALYSIS?

From a management perspective, there are several reasons for using cost-benefit analysis. First, if the agency's programs are amenable to cost-benefit analysis, the techniques can help the agency allocate its resources among programs. If

Programs A and B have the same costs but A generates more benefits than B, the efficiency-oriented agency will choose A. A variety of rational management techniques, moreover, implies selecting the program alternative that returns the greatest benefits. These include program-planning budgetary systems (PPBS), zero-base budgeting, and strategic program planning.

Second, even if calculation of costs and benefits is extremely difficult, approaching programs from a cost-benefit perspective is valuable. In examining costs and benefits, the manager focuses on the goals of the program and thus sharpens his or her ideas about what is to be accomplished. The cost-benefit approach also can bring to the manager's attention program features about which he or she may not have been previously aware. Often an analysis will conclude that some particular benefit cannot be measured or estimated. By pointing out areas of ignorance, cost-benefit approaches raise red flags that warn managers about areas that need further study or areas where managerial judgment must be exercised.

Third, legislation often requires the use of cost-benefit analysis. For example, the National Environmental Protection Act (NEPA) of 1969 encouraged the use of cost-benefit analysis in environmental policy, and the Flood Control Act of 1936 specified that, for water projects, a comparison of benefits and costs be made. More recently, President Ronald Reagan's Executive Order 12044 required that benefits exceed costs before regulatory rules are issued. Given the fiscal constraints on government today, pressures for greater efficiency and for greater use of cost-benefit analysis will increase.

Fourth, the manager should be aware that cost-benefit analysis has political as well as managerial uses. The Army Corps of Engineers is thought to have used favorable cost-benefit comparisons to gain authorizations for questionable projects. For example, the recreation benefits for a flood control project might not take into account the recreation services provided by previous projects. Other people believe that cost-benefit analysis was used in the 1980s to prevent agencies from acting, thus creating paralysis by analysis.[1]

ASSUMPTIONS OF COST-BENEFIT ANALYSIS

Cost-benefit analysis makes certain assumptions about the program that is being analyzed and the world around it. To the extent that these assumptions do not correspond to the assumptions held by the manager, cost-benefit analysis is an

[1]See Kenneth J. Meier, "The Limits of Cost-Benefit Analysis," in *Decision making for public administrators,* ed. Lloyd G. Nigro (New York: Marcel Dekker, 1984), pp. 43–46.

inappropriate technique. Every manager who receives or manages a cost-benefit study, therefore, must question the assumptions underlying the study. Failure to examine the assumptions means that the manager tacitly accepts them.

IMPACTS

Cost-benefit analysis assumes that all the major or important impacts (costs or benefits) of a project can be identified or described. For projects that do not have clear goals (whether by legislative intent or for other reasons), cost-benefit analysis may offer analytical solutions that are too easy. For programs where the causal linkages and thus the impacts are unclear (for example, the impact of environmental regulation on general health), the cost-benefit analysis also will be unclear. The more confident the manager is that all program impacts can be identified, the more confidence the manager can have in the analysis.

QUANTIFICATION

Cost-benefit analysis assumes that every impact can be measured in a common unit of measure, usually money, so that alternatives can be prepared. When it comes to unmeasurable items, a cost-benefit analysis is no better than any other quantitative technique. It cannot compare dollars saved with air cleaned, but it can compare dollar savings as the result of cleaner air with dollars spent to clean the air. Again, if the manager is confident that all the major costs and benefits can be measured in a common unit of measure, the manager may have confidence in cost-benefit analysis.

INDIVIDUAL KNOWLEDGE

Cost-benefit analysis assumes that each individual is the best judge of his or her own self-interest.[2] Individuals' assessments of the value of the project to them help determine the benefits and costs of a project. For example, the benefits of a waterfowl refuge are assumed to be the sum of the values that all citizens, acting as individuals, place on the refuge. Cost-benefit analysis assumes that individuals know the value of program consequences in both the present and the future.

---

[2] Edith Stokey and Richard Zeckhauser, *A primer for policy analysis* (New York: Norton, 1978).

MAXIMIZATION OF THE DIFFERENCE

Cost-benefit analysis assumes that the goal of a program is to maximize the difference between benefits and costs. Although this seems like an obvious assumption, in many cases public policy programs are not intended to maximize the benefit/cost ratio. For example, a legislature might prefer a program that produces a lower benefit/cost ratio but provides more benefits to the poor. An administrator concerned with program processes might reject a welfare fraud program that has high benefits compared with costs if the procedures used are unacceptable. Public programs have numerous goals, and maximizing benefits in relation to costs may or may not be one of them.

THE TECHNIQUES OF COST-BENEFIT ANALYSIS

On a theoretical level, the techniques of cost-benefit analysis are straightforward and attractive to the decision maker seeking to allocate agency resources efficiently. The steps in a cost-benefit analysis are as follows:

1. Identify the project or projects under consideration as specifically as possible.
2. List all the impacts, both negative and positive, on society both in the present and in the future.
3. Provide a monetary estimate for each identified impact—either positive (benefits) or negative (costs).
4. Calculate the net benefits for the project or for each of the alternative projects by subtracting the total costs per project from the total benefits per project.
5. Make a choice based on the decision criteria that have been established, or present the information to the designated decision maker clearly enough that he or she can make a decision.

So far, it looks simple, almost mechanical, but applying these steps to a project or group of projects will raise a number of questions that require a good deal of thought and resourcefulness on the part of the evaluator. The evaluator must therefore have a firm understanding of the decision constraints (such as money, policy alternatives, and so on) that are operative for that project and the economic concepts that are relevant for the type of measurements being attempted. Each of these steps will be discussed in turn.

IDENTIFYING THE PROJECT

Any manager commissioning a cost-benefit study must define carefully the project to be studied. For example, a study examining "all the welfare programs of the state of Oregon" to measure their benefits and costs is so vague that it is meaningless. Does such an analysis include just traditional welfare programs, or does it include related programs such as vocational rehabilitation and veterans' preferences? Do programs that contribute to maintaining the well-being of temporarily disadvantaged citizens, that are not normally considered welfare (for example, unemployment), come under this definition? Definitions that are too narrow should also be avoided. Doing a cost-benefit analysis of placing sand and water cushions in front of barriers on highways without considering the other improvements in safer roadways will result in an analysis that cannot distinguish the benefits of the cushions from those of raised traffic lane bumps, breakaway posts, limited-access roads, and divided highways.

Identifying a project precisely cannot be done by formula. At the very least, project identification requires attention to legislative intent, administrative promulgations, and agency implementation procedures. Perhaps the most effective method of identifying a program for cost-benefit analysis is a consensus-building group process involving program managers to identify the crucial aspects of the program. No matter how the identification is done, every cost-benefit analysis (and every program evaluation) should contain, at the beginning of the report, a description of the crucial elements of the program (see the discussion of goal-setting in Chapter 5).

LISTING THE IMPACTS

Once a project has been defined, the analyst begins to compile a list of all the major impacts (costs and benefits) of the proposed project. Two major problems should be tackled at this stage of cost-benefit analysis. The first is where to find the information: What sources and methods can help anticipate and forecast impacts? The second is how to classify the impacts: What schemes will help the analyst avoid "double counting" impacts while providing a way to separate the impacts into meaningful categories?

COLLECTING DATA

Collecting data on potential impacts is one of the most difficult and time-consuming aspects of a cost-benefit analysis. The type of data gathered depends to some extent on the levels of expertise of program staff and the amount of money and time available for the study. Wherever possible, impacts should be identified both by experts and by interested parties, whether citizens or special interest groups. Methods for collecting data include using staff experts or contracting with outside experts, literature searches, involving experts and citizens through advisory committees, conducting surveys, and brainstorming with colleagues or with decision makers. Data sources that provide clues for potential impacts and trend information needed for forecasting include statistical reports prepared by various levels of government, census materials, and reports on similar projects. An expensive way to obtain information on possible impacts is to conduct a pilot study, but funding and time for this type of preliminary effort is rarely available. Besides, in order to maximize the opportunity for direct observation of impacts and parameter changes, a good project evaluator would need to identify as many potential impacts as possible before setting up such a pilot study. Another way to generate information on impacts is to set up a simulation using several of the project choices in order to determine how they might affect several key variables or parameters. The problem with this approach is that the analyst usually needs quite a bit of background information about the important impacts to set up the simulation.

CLASSIFICATION OF DATA

In addition to determining where and to whom to look for the information, the analyst should set up a systematic approach to classifying each identified impact. Preestablished categories can be confusing because some impacts do not fit neatly into categories. A well-thought-out classification system can prevent "double counting," one of the major problems with early efforts in cost-benefit analysis. Double counting usually occurs when the analyst includes the increase in the price of a good as a cost and then also includes the effect that price change has on a group of people as a cost. For example, if a new shopping center in a section of the city results in increased property values for houses in that area, the analyst should not count this increase in property values as a benefit. The increase in property values represents the additional convenience and opportunities for jobs, entertainment, and shopping that are important to consumers, and are the benefits that should be incorporated into the cost-benefit analysis. Property values and enhanced opportunities cannot *both* be included in the analysis because they represent a stream of effects rather than two separate benefits.

There are three major ways to classify costs and benefits: by internal and external effects, by tangible and intangible effects, and by direct and indirect effects.

The important distinction between *internal and external effects* is receiving even more attention as a critical issue in environmental programs. Internal effects are those that are specifically linked with the project to be undertaken and the goals and objectives that have been identified as defining the project. External effects are those that fall outside the definition of the project goals and objectives and thus are usually not given a monetary value in the analysis. The decline in the violent crime rate as the result of a successful drug-abuse treatment program is an example of an external effect. Often called second-order consequences, external effects are excluded from many analyses but, because the second-order benefits and costs of a program may well dwarf the direct benefits and costs, they should not be excluded. If the analysis is to reflect costs and benefits accurately, external factors must be considered. Analysts could specify program benefits and costs ad infinitum, but at some point they must assume that all other benefits and costs are trivial or cancel each other out. Finally, the manager should make sure he or she accepts the assumptions of the analyst.

Increases or decreases in production or consumption opportunities for the public can be designated as *real benefits* or *real costs.* This is in contrast with *pecuniary benefits* or *pecuniary costs.* Pecuniary effects change the financial outcomes or effects for the public or society without specifically affecting individuals. Thus, the increase in property values resulting from building a new recreation area would be a pecuniary effect and would not be counted, while the actual changes in leisure-time consumption in the neighborhood resulting from the convenience of the recreation area and the increased opportunity for play would be counted because they are real benefits affecting consumption opportunities for the public.

Tangible and intangible effects are terms frequently used by groups that disagree with the values placed on impacts in a cost-benefit analysis. Tangible effects can be easily assigned a monetary value based on market information. Intangible effects are effects for which the market does not supply good information. Some analysts break this down further, using the term *incommensurables* to describe effects that, while difficult to quantify and measure, can be valued in at least a relative measure. The analysts then reserve the term *intangible effects* for effects that are not economic in nature, such as saving a human life or preserving democratic processes. Some contend that there are no intangible effects, that by using one technique or another all effects can be treated as tangible.

Identification of *internal and external effects* (costs and benefits) should follow certain steps. First, all internal costs (for example, operating costs of the program and fixed overhead costs) should be estimated. Second, any external project costs (for example, regulations compliance costs, increased noise pollution) must be identified. Third, direct benefits are to be specified: What are the direct program benefits intended by the agency? Fourth, external benefits (for example, positive second-order consequences) should be outlined. Fifth, each of the benefits and costs must be examined to determine whether it is a real effect or a pecuniary effect; pecuniary effects should be deleted from the analysis, real effects should be retained. Sixth, the manager must determine whether the benefits and costs are tangible or intangible. This final determination begins the next step in the process—assigning monetary values to program impacts. Some analysts omit either intangible impacts, or minor impacts, or both. Whenever impacts are excluded from the analysis, the report should specify what effects are excluded.

MAKING MONETARY ESTIMATES

Cost-benefit analysis, especially the process for assigning monetary values to program impacts, is derived from a branch of economics called welfare economics. Welfare economics assumes that the welfare of a society depends on the well-being of the individuals in that society and nothing more. This means that when considering policy choices the analyst must be concerned with the value that individuals place on the consequences or impacts of a policy. The "social welfare" can then be determined by adding up, or summing, all the individual statements of value. This is a sound idea if one can accept the assumption that individuals can accurately place values on program effects. Although this assumption fits well with democratic principles, recent threats to the environment and to personal safety indicate that individuals can be misled by lack of good information regarding policy consequences. Moreover, the values some individuals place on program effects can be unduly affected by such problems as deviance or envy, which cause them to choose options that may be personally harmful or against the general well-being of society. Examples of this type of deviant counterdecision are failure to use seat belts, cigarette smoking, building homes in known floodplains, and buying expensive items to keep up with the Joneses.

PRICING TECHNIQUES

Although welfare economists have never successfully measured the benefits an individual receives from a project, they have generally agreed on a substitute—voluntary transactions. If Citizen A voluntarily buys Good B from Citizen C for five dollars, it is obvious that Citizen A believes that Good B is worth at least five dollars; otherwise, A would not make the transaction. Willingness to pay therefore becomes the general criterion for establishing value. This assumes that rational individuals seek more rather than less of a desired good, and seek less rather than more of an undesired good.

Market Prices. Market price is the first technique used for estimating value. By using market prices to measure individuals' values–and thus social values–economists must assume that the market is a perfect reflection of supply and demand. As we will see, this assumption is often troublesome. However, both because some segments of the market are still near ideal and because there is no other information available, even the most thoughtful critics tend to accept such calculations, albeit with reservations.

The analyst measures social value by using market information on the price and quantity of a good or service in order to put together the familiar demand curves (see Figure 3-1). The logic underlying the use of demand curves is that *rational* people seek to maximize their personal wellbeing to the fullest. In a perfect marketplace, therefore, the price that the rational person would be willing to pay for a product would be exactly equal to its price. This price would be at point *B* on the demand curve line in Figure 3-1. From this, economists conclude that if a project produces a relatively small increase or decrease in the quantity of the good or service available on the market, the change can be valued as equal to the quantity of change multiplied by the market price. The social value of a project, then, is simply the sum of these changes in a good or service multiplied by the price of the good or service. Since a market demand curve represents the total of all individual demand curves, it tells the analyst how much of a product would be desired at a certain price by the public in general, thus excluding consideration of which individuals were doing the buying.

For example, the National Transportation Safety Board requires that all automobile bumpers be able to withstand a five-mile-an-hour collision, with no resulting damage to the vehicle. One benefit of this regulation is that fewer vehicles are damaged in low-speed accidents. The market price value of such an accident being avoided is the cost of repairing the vehicle. The total of this

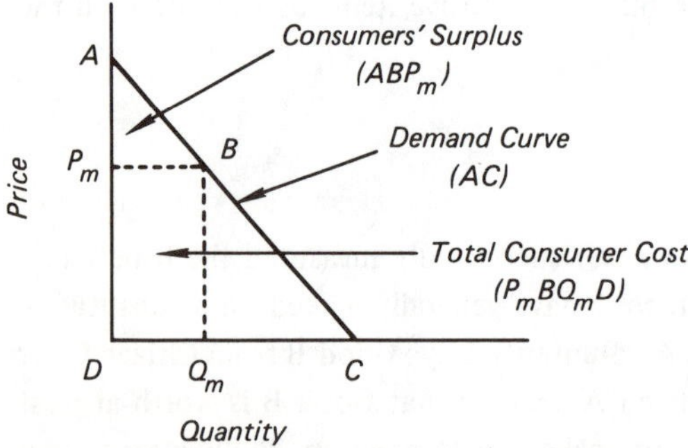

FIGURE 3-1
A typical price-quality-demand curve

benefit is therefore the product of the number of accidents without damage and the cost of repairs had the accidents occurred.

Shadow Prices. In some cases market prices do not accurately reflect social value because the market price is too high or too low. Recent cost-benefit analyses often include values other than market prices to reflect social value. These adjusted values are known as shadow prices. The rationale for computing shadow prices is that, in situations where market prices are not available or are not appropriate, the analyst must substitute a value (shadow price) that explicitly values the benefits and costs in question. The manager must realize that because all shadow prices are based on the subjective analysis of the analyst they may be controversial and can undermine the credibility of the analysis. The procedures used for and the assumptions underlying calculation of shadow prices should be stated by the analyst in the report. There are two instances where shadow prices are valuable in analysis: When market prices exist but seem inappropriate or are clearly biased (for example, in cases of monopolies) and when market prices do not exist (for example, where human safety is at stake).

Biased Market Prices. Market prices accurately reflect social value and people's willingness to pay when the market exhibits characteristics of a purely competitive market. In a purely competitive market there are so many buyers and sellers that no single person can affect the market price, all goods are equally substitutable, consumers have perfect information, and producers have freedom of entry and exit. Whenever the market deviates from perfect competition, market prices deviate from social value.

In monopoly situations, for example, one seller determines the market price of a good. In this case the seller can and understandably will restrict production and charge prices greater than the "normal" market price. The more people desire the product and are unwilling to do without it, the higher the price the monopolist can charge. In an oligopoly situation, the results are similar but not as extreme. While all economists agree that monopoly and oligopoly prices are too high, they do not agree on how to adjust the prices to reflect social value. There are, however, ways to approximate market prices.

Determining the social value is of critical importance when the price of the goods in question changes dramatically without a corresponding change in the quantity available. Instead of using the quantity-to-price ratio to determine the net benefit produced by such a change, the analyst should use the change in the consumer's surplus to determine the net benefit to the consumer. The *consumer's surplus* is the excess amount of utility that individuals obtain because they do not have to pay as much as they would be willing to pay to buy the quantity desired (see Figure 3-2). For example, if completing a project will decrease the

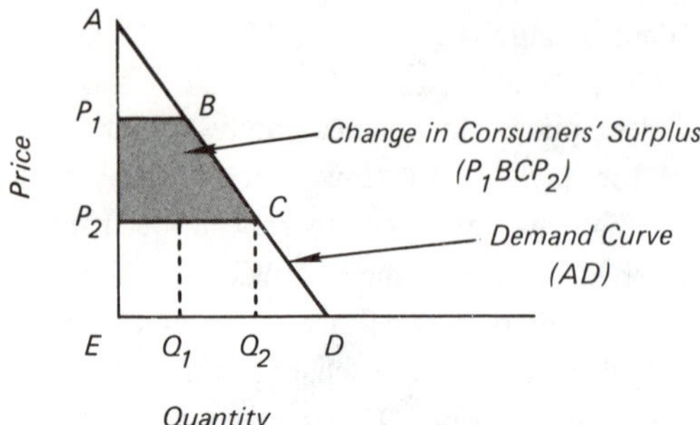

FIGURE 3-2
The relationship of the consumer surplus to normal price/quantity ratios

price of water to consumers (from P_1 to P_2), consumers will get a surplus or extra benefit represented by the shaded area in Figure 3-2.

Market prices can also be too low. For example, if government subsidizes the price of a good, the market price will be too low. Gasoline prices in the 1960s were subsidized, and today grain prices and food prices in many other countries are subsidized. The cost-benefit analysis must adjust the value of these goods upward. Again, there is unfortunately no accepted way of making this adjustment.

Other market conditions that can cause deviations from normal price/quantity ratios are government actions in the form of taxation, regulation, and price controls. These can generally be incorporated into the market price analysis by making explicit the extent of the price differential but using the market price nonetheless. Price controls, for example, bias the market by setting artificial floors or ceilings on goods, thus creating a gap between the market price and the social value of such goods. The actual effects of these government actions usually depend on the supply conditions; such action does not always create a discrepancy between market prices and social values.

The problems with creating shadow prices raise the question of whether the analyst should even try to correct for market information. Although the unnecessary use of shadow prices should be avoided, the analyst must, nonetheless, be alert to markets that are clearly biased. In such cases, the analyst must supply information on deviations from social value. Other factors in the evaluation may also be undervalued or overvalued if they are rated at their actual market prices. For example, inputs to production such as labor costs are often biased in project evaluation estimates. If a project creates jobs, these new jobs are benefits only if the jobholders would otherwise be unemployed. Even if the jobholders were unemployed and hired, thus creating a benefit, they were not likely to have remained unemployed over the whole period of the project (unemployment seems to be cyclical), so labor costs must reflect job market conditions during the project. In a similar manner, capital and land that would otherwise not be used during the life of the project should be valued at zero social cost when included in a project evaluation.

From the manager's perspective, shadow prices are easier to manage. The manager must insist that each use of shadow prices be documented and fully justified and then decide whether the shadow prices are consistent with the manager's own view of the program and its environment.

Absence of Market Prices. In some situations, market prices, even biased ones, simply do not exist. Shadow prices are frequently used in two situations without market prices–externalities and public goods.

Externalities are effects that involuntarily accrue to outsiders. Air pollution is a classic example of an externality. The analyst must decide which external effects are to be included, remembering that real changes in production or consumption opportunities should be included, while pecuniary changes affecting outsiders are usually excluded. In all cases, there is no compensation for costs borne by outsiders and there are no fees collected for benefits enjoyed by outsiders.

Putting a price on noise is the most frequently cited example of how difficult it is to create market prices for external effects. While analysts can measure the amount of noise or level of noise with relative ease, they have enormous difficulty putting a price on each unit (decibel) of noise. Conducting a survey would be one solution, although the reliability of individuals' statements of willingness to pay for quiet is likely to be low because the absence of noise is an issue that falls under the realm of public goods. In fact, the issue is really one of compensation rather than willingness to pay. That is, if peace and quiet are a right under government protection, the proper concern for the analyst is how much money a citizen would accept to give up the right to peace and quiet, or how much money would compensate for the loss of peace and quiet. The problems in pricing externalities are similar to those in pricing incommensurables and will be discussed shortly.

The issue of *public goods* creates additional problems. Public goods are goods provided or protected by the government and consumed by many people at the same time, for example, clean air, mosquito control, national defense, wilderness areas, and highways. These types of goods and services can be characterized in three ways:

1. The use of a public good by one or more individuals does not affect its availability to others.
2. Public goods are in theory nonexclusive in that, once the public good is provided, citizens are not selectively excluded.

3. Public goods will not be provided by individuals if it is left strictly up to them, even though such goods provide a net benefit to society as a whole.

Try as they might, economists have been unable, as yet, to develop a model that can be applied to valuing public goods. The reason for this is the individual's alleged rational behavior. Theoretically, citizens use a gaming approach when asked how much they are willing to pay for a good. By undervaluing their interest in a public good, individuals hope to be able to take advantage of the good without having to take any responsibility for providing it. Thus, individuals act as "free riders" by consistently underestimating their willingness to pay for public good.

Faced with this problem, analysts have several less-than-ideal choices. They can take a survey, knowing that the information obtained will be severely biased. They can substitute the market price of a similar private good (for example, using the price of books to determine value of library circulation), but this would have to be well documented through analogous reasoning. They can use a referendum in which people vote to tax themselves to provide a benefit, an alternative that has received increasing support from citizens. The feasibility of using a referendum for any project evaluation would depend on the political unit, citizen interest, and so forth.

In summary, shadow pricing is a technique for uncovering deviations from market prices that are important enough to be presented to the decision maker as evidence of bias in the final cost-benefit calculus. Except in cases of externalities and public goods, shadow prices should be used judiciously in altering the market prices used in the cost-benefit analysis. All shadow pricing should be well-documented both to justify the decisions made by the analyst and to inform the manager of the extent and direction of the biases in market prices. The issue of shadow pricing is controversial. Many critics believe that these biases must be included in the actual cost-benefit calculation. They point out that managers are not apt to read the fine print and tend to rely too heavily on the numbers provided in the final tally, overlooking the assumptions made and the cautionary statements offered. And critics claim that, because no analyst is value-neutral, the analyst who believes a market price is severely biased has an obligation to document the bias and adjust the social value included in the analysis.

Incommensurables. An even more controversial area for cost-benefit analysis is the quantification of incommensurables. Incommensurables are effects for which no market exists. Human lives, health, clean air, and recreational opportunities are some goods that might be considered incommensurables. In new areas of social regulation, such as worker health and safety, environmental protection, and consumer protection, the major benefits are incommensurables. As a result, cost-benefit analysis must face the problems in valuing incommensurables if it is to play a role in these growing areas of government action.

The best example of problems in valuing incommensurables is the area of human lives. Government policies that save lives (traffic safety, worker safety, and the like) do provide some benefits. Analysts who accept the principle of willingness to pay might logically attempt to determine how much compensation a person would want before that person would give up his or her life. Of course, willingness to pay would not yield an appropriate monetary value for a human life because most people are unwilling to accept any amount of monetary compensation in exchange for their lives.

Often willingness to pay measures are constructed by reversing the estimation process. Rather than asking individuals how much compensation they require, an analyst asks how much an individual would be willing to pay to avoid a cost or to attain a benefit. For example, an analyst might ask a citizen how much he or she would be willing to pay in taxes to eliminate noise from a nearby airport. While this approach works for some effects, it has serious problems when applied to valuing lives or injuries. Each individual's willingness to pay is limited by that person's income, thus giving greater weight to the preferences of wealthy citizens than to those of poorer citizens. This violates both the norms of democracy and the principles of welfare economics.[3]

Another approach used to estimate the value of saving a life is the proportionate risk approach. Individuals are asked how much compensation they would accept to be subjected to a 1 percent greater risk of death. In industrial safety, for example, a person might be told that Job A pays $5 per hour and that Job B has a 1 percent greater risk of on-the-job death, and then be asked how much the wages should be for Job B. Let us assume that the answer is $1 more per hour, or $2,080 per year. The analyst then assumes that if 100 workers are exposed to this greater risk, one will die during the year. The required compensation for one life, therefore, is $2,080 × 100 = $208,000. The manager should be aware that the proportionate risk approach has a few problems. Getting survey responses from workers on such a question is extremely difficult. In addition, translating the proportionate risk to an individual into an actual number of deaths may well change the workers' calculus.

Another method often used to place a value on an individual's life is the *consumption method.* By assuming that an individual's contribution to society can be measured by what the individual produces, the cost-benefit analyst then estimates the value of an individual's life as the amount of money that individual would earn for the rest of his or her life (the insurance industry uses the figure

---

[3] I. M. D. Little, *A critique of welfare economics* (London: Oxford University Press, 1957).

$200,000 in 1972 dollars). Under this approach human beings are treated exactly like government programs and valued according to the future income/ benefits they produce. The consumption approach to valuing human lives creates the greatest ethical problems for the manager because it places a higher value on men than on women, on Whites than on Blacks, and on the able-bodied than on the handicapped. The criticism of analysts who use this approach is extremely difficult to counter.

The above methods are only a few of the ways that can be used to value human lives. Mark Thompson discusses twelve different approaches to estimating the value of a human life, each based on different assumptions.[4] An easy, noncontroversial way of valuing lives is not likely to be found soon.

Even though valuing loss of life or injury is an enormous problem in cost-benefit analysis, such values are critical in producing studies that evaluate the impacts of the expanding technological sector of our society. Conditions of risk and uncertainty dominate such evaluations and create dilemmas both for the experts trying to assess the impacts and for the decision makers trying to use such assessments. The issues that have to do with valuing such impacts are so deeply embedded in philosophical and/or religious traditions that the credibility and legitimacy of such estimates will continue to be debated.

DISCOUNTING THE VALUES

Estimating the values of effects cannot stop when values are determined by market prices, shadow prices, or some other technique. Few projects result in costs and benefits that occur immediately after implementation; most have impacts occurring in different years over a period of time following implementation. Cost-benefit analysis uses discounting to incorporate the effect of time on both the costs and the benefits.

Discounting is based on the theory that people will not pay as much for something that will not be available until a future date. In other words, people prefer to sacrifice a benefit in the future in order to have a benefit in the present. The discount rate has an enormous impact on the net present value. If the discount rate is set at 5 percent, a person who has been promised a benefit of $100 from a project next year can calculate that the $100 in today's value is worth only $95.24. The formula for this calculation is:

$$PV = \frac{FV}{(1+d)^t}$$

---

[4] Mark S. Thompson, *Benefit-cost analysis for program evaluation* (Beverly Hills, Calif.: Sage Publications, 1982).

where *PV* is the present value of the future benefit, *FV* is the stated value of the future benefit, *d* is the discount rate, and *t* is the number of years. In the case of the person who was offered $100 next year, the present value would be determined as follows:

$$PV = \frac{100}{(1+.05)^1}$$

$$= \$95.24$$

The discount rate is helpful because it reduces items that cannot be compared—costs and benefits occurring in different years over a time period—into a common monetary unit. This unit is the present values of the costs and benefits regardless of when they are predicted to occur. Discount rates look like and function like interest rates.

At least two major problems or criticisms have been raised about the use of discount rates. The first does not quarrel with the use of discount rates so much as it simply points to the extreme sensitivity of final benefits and/or costs to the discount rate chosen. Even a variation of one or two percent in the discount rate can make the difference between one project and another. This leads to the second problem. The appropriate rate to use is the one that produces the actual present social values for the project. This appropriate rate is known as the *social discount rate,* which should be an accurate measure of the rate at which society will trade off present costs and benefits for future costs and benefits. Finding this rate and getting agreement on it is very difficult. A number of sources exist for choosing the rate, and no one rate seems to have proven itself as *the* social discount rate.

Sources of Discount Rates. Examples of the rates that have been proposed for use as the social discount rate are: market interest rates usually associated with government bonds, banks, or savings and loan associations; marginal productivity of investment; corporate discount rates; the government borrowing rate; personal discount rates; and personal preferences for social discount rates. Each of these examples has further nuances—for example, there are usually several different market interest rates available, depending on the amount of risk to the lender. Economists and policy analysts have their favorite social discount rates, the justification for which is usually based on a belief in either high or low discount rates.

Low or High Rates. Low discount rates favor projects whose benefits occur further in the future. On the basis of work by economist A. C. Pigou, proponents of a low discount rate argue that because individuals selfishly weigh their own interests too heavily compared with those of future generations, the government must intervene to correct their shortsightedness by acting as a trustee on behalf of future generations through the use of lower discount rates which make long-term projects more feasible. Those arguing for a high discount rate state that in order to get a meaningful analysis the true opportunity costs of a project must be considered. They point out that the higher rate will knock out many long-term, capital-intensive projects, freeing money for private business forays into the economy and for more short-term public projects.

Problems with Discounting. The most basic question is whether discounting makes any sense in evaluating public projects. Particularly in safety and environmental issues, such as nuclear power development or wilderness preservation, critics have made a strong case for the folly of discounting or weighting the future less heavily than the present. Besides, they point out, the low discount rates used cannot handle benefits and costs exceeding about fifty years.

Table 3-1 illustrates the present value of $10 at the end of specified years for several discount rates. The low range of present values at higher discount rates (10%-15%) and for longer periods of time (twenty-five years and over) demonstrates why analysts are wary of discounting the distant future heavily, especially for so-called "second generation" impacts, and seldom use discount rates beyond fifty years. The analyst is really on the spot here. Economists have not been able to agree on the social discount rate but the federal government, in some cases, has made the decision for the analyst by recommending, in years past, two discount rates to be used in evaluations by or for its agencies—7 percent for water projects and 10 percent for all others. Currently, the discount rates for government programs are tied to the Treasury Department's long-term borrowing rate, which fluctuates, making the discount rates better reflections of the true cost of investments in public projects. Discount rates for water projects are slowly being brought into line with other projects. Discounts for these projects may not increase more than ¼ of 1 percent per year. Thus, in 1984, the

TABLE 3-1 Present value of $10 at the end of year (*t*) under selected discount rates (*d*)

	Discount rates (*d*)				
Year (t)	*5%*	*7%*	*10%*	*12%*	*15%*
1	9.52	9.35	9.09	8.93	8.70
5	7.81	7.14	6.21	5.68	4.98
10	6.13	5.08	3.86	3.23	2.47
15	4.81	3.62	2.39	1.83	1.23
20	3.77	2.58	1.49	1.07	.61
25	2.95	1.84	.92	.59	.30
30	2.31	1.31	.57	.33	.15
40	1.42	.67	.22	.11	.04
50	.87	.34	.09	.03	.009

discount rate for water projects was $8^{3}/_{8}$ percent while other projects vary between eleven and thirteen percent. This solves the problem for the analyst, but not necessarily for the manager.

COMPARING THE COSTS AND BENEFITS

THE NET PRESENT VALUE

Once important program effects have been given a monetary value and discounted, the next step is to compare them.

Net present value (NPV) is the method of comparison most often used to evaluate the alternatives. It combines a number of factors that influence the real monetary value of a project. By discounting any costs and benefits that will occur in the future, NPV considers the problem of time in evaluating the current value of the project to society. The formula for calculating the NPV would be

$$\frac{B_t - C_t}{(1+d)^t} \;\ldots\ldots + \frac{B_n - C_n}{(1+d)^n}$$

where B_t is the monetary value of benefits at time t, C_t is the monetary value of costs at time t, d is the discount rate, and n is the number of years of the project's life. This means, for example, that the NPV of a two-year project that yields a net benefit of \$100 the first year and a net benefit of \$50 the second year would not be evaluated as worth \$150. Using the appropriate discount rate (5 percent for example), the NPV of such a project would be \$140:

$$\begin{aligned} \text{NPV} &= \frac{100}{(1+.05)^1} + \frac{50}{(1+.05)^2} \\ &= 95 + 45 \\ &= 140 \end{aligned}$$

If the analyst has already discounted each of the individual effects, the NPV formula simply translates into total discounted benefits minus total discounted costs. If the NPV is positive, a project returns benefits in excess of costs; if it is negative, it does not. When comparing two projects, the one with the highest net present value returns the most benefits over its costs.

Net present value is not the only bottom-line figure that the manager can see in a cost-benefit analysis. Other criteria are sometimes used but not as often as net present value.

CUTOFF PERIOD AND PAYBACK PERIOD

A cutoff period is often used in private industries that take calculated risks but do not wish to overextend their commitments. Establishing a cutoff period of ten years, for example, would mean that no benefits or costs (usually just benefits) that accrue after ten years will be counted. Cutoff periods are not often used in the public sector because government seldom needs to worry about such risks and time constraints. The cutoff criterion also favors projects yielding quick returns on investments and discriminates against programs whose benefits occur further down the line. The payback period is the length of time it takes for a project's benefits to equal its costs. This criterion suggests that a project that recovers its initial costs in the shortest period of time is the most desirable project. It favors projects with immediate payoffs and discriminates against those with large but long-term benefits.

INTERNAL RATE OF RETURN

The internal rate of return is determined by increasing the discount rate until benefits are equal to costs; the discount rate that equalizes benefits and costs is the internal rate of return because it specifies the annual percentage return on investment yielded by the project. Until recently the internal rate of return (IRR) was used extensively. It is now criticized primarily because of difficulties in interpreting the information it provides. The manager should undertake only those projects that have an IRR greater than the established discount rate. The logic is that it pays to invest public funds in a project that earns returns at a higher rate than the government could borrow funds. If a comparison is being made among several alternatives, the decision maker would look for the project that yields the highest IRR. The IRR criterion, however, is best applied only to projects that require a one-time, critical investment, with returns or benefits flowing over a period of time. Critics point out that calculating an IRR would not be as appropriate as calculating the NPV in instances when the project's discount rate changes in midstream, when costs are spread throughout the life of the project, or when the budget is constrained in some manner.

BENEFIT/COST RATIO

The benefit/cost ratio results from dividing the net discounted benefits by the net discounted costs. This may seem to be a simple criterion, and it has been popular in the past, especially for demonstrating the benefits of water projects throughout the United States. The major flaw in using this ratio as a sole decision rule is that a project that has the highest benefit/cost ratio may actually have the lowest total net benefits of all the projects under consideration. People may want to spend more money proportionately in order to gain the higher net benefits provided by one of the other projects. As a general rule, the benefit/cost ratio is most useful in situations where the amount of money to be invested in a project or group of projects is limited.

EQUITY

The last decision criterion is equity. Equity is not always considered in cost-benefit analysis but recent pressures from interest groups as well as from evaluators who are aware of the need to recognize democratic principles and normative issues in their appraisals have led to a movement to include equity as a separate issue in presenting these analyses. No effective way of incorporating equity in the mathematical models of the economists has been suggested. The general approach at this time is to prepare a supplementary report (or a distinct section in the report) dealing with the issues of the distribution of both benefits and costs, drawing attention to special burdens or benefits adhering to a group or several groups. This reserves the equity considerations for the manager. The inability of cost-benefit analysis to pick up the equity issue as a central part of the formal analysis has come under serious attack. Critics claim that even when a special report on equity considerations is included, the issue is not given full consideration by managers, primarily because such problems are too difficult to include in the decision calculus without relying on intuition or subjective judgment. A manager sensitive to equity concerns must therefore specifically consider the distribution of costs and benefits when making a decision.

SENSITIVITY ANALYSIS

The determination of net benefits is often the result of some numbers in which the manager has only a little faith (for example, the value placed on a life, or an arbitrary discount rate). In such cases a technique known as sensitivity analysis can address the concerns of the manager. Sensitivity analysis simply varies the value of one effect (say, the discount rate) until net benefits are equal to zero. Any discount rate below the final rate determined by the sensitivity analysis would result in positive net benefits. In this way, sensitivity analysis allows the manager to determine how crucial some of the assumptions and values are to the final results.

MAKING CHOICES BASED ON COST-BENEFIT ANALYSIS

The major contribution of the cost-benefit analysis is its systematic approach to planning and evaluation. If the analyst plans the project evaluation carefully and remains alert to the potential weaknesses of this type of analysis, the final product can contribute to the decision-making process. The cost-benefit specialist should remember that the technique is designed to organize information for the decision makers. The analyst's task is to see that the information is of the best quality, given the constraints under which he or she is operating, and that any valuing done is set forth as explicitly as possible.

IDENTIFYING AND DESCRIBING THE PROBLEM

The primary responsibility for defining the problem or project to be evaluated rests with the manager for whom the study is being done. However, the analyst should, if possible, work closely with the manager at this stage to be sure that the analysis is designed and focused to evaluate the problems and exact goals and objectives in order to provide information that will be useful to the decision maker. If this is accomplished, the analyst will be more likely to produce a cost-benefit analysis that deals with problems relevant to the policy process. Every problem has unique features, which should be identified and described early in the planning stages. Tasks that must be undertaken include (1) drawing up a description of the status quo—that is, what the effects of the current policy are or would be; (2) identifying constraints that may impede implementation of project goals and objectives (for example, budgetary or resource limits, legal obstacles, administrative or institutional limitations, political problems, social considerations, and technological constraints); and (3) choosing decision criteria, such as the discount rate, the relevant time period, and the variables to be investigated.

SETTING UP THE DESIGN

Ideally, the cost-benefit analysis would systematically cover all aspects of a problem, but constraints on time, resources, and staff mean that the design must focus on some effects more than on others. In addition, the measurement and analysis of data will have to be based on the expertise available.

COLLECTING AND ANALYZING THE DATA

The format for presenting the data analysis should be determined before data collection begins. Original data should be collected when feasible, and sources used should be recorded for later updating and documentation. Both quantitative and qualitative data have a role in cost-benefit analysis. The analysis and interpretation of each should be carefully planned, documented, and justified.

PRESENTING THE RESULTS

The rationale for performing a cost-benefit analysis is to inform managers in order to improve public policy. If the cost-benefit analysis is well planned and executed, the manager will have been kept informed of the progress through preliminary work plans and summary reports. Because cost-benefit analysis is controversial, the assumptions made in the analysis must be made explicit for those who will be basing their decisions on the final report. And, where possible, the report should include a discussion of other points of view that were not incorporated in the analysis. In addition, the manager must be sure that the study is written in prose that lay people can understand rather than in the equations of the economist.

THE PROS AND CONS OF COST-BENEFIT ANALYSIS

The potential of cost-benefit analysis as an analytical tool for decision making on public policies and projects is more controversial now than in the past. Supporters give it high marks for the order it brings to problem solving, the way it opens the governmental process to public scrutiny, and the information it consolidates.

But critics point to the theoretical and technical difficulties in applying cost-benefit analysis in the public sector. They claim that because it ignores the political process, critical questions raised by democratic principles may go unanswered, even when such questions as equity or conflicting values are addressed. According to these critics, cost-benefit analysis encourages decision makers to put too much weight on quantitative information. Considerable attention has been focused on the problem of substituting the judgment of "experts"—whether engineers, scientists, systems analysts, or economists—for the judgments of the public. As a result, some efforts are now being directed toward subordinating cost-benefit analysis to the political process through renewed emphasis on public participation, consensus building, and the peer review process. Regardless of the outcome of this dispute, the manager must always subordinate cost-benefit analysis to management needs.

The rules regarding when cost-benefit analysis is the best choice for studying policy alternatives are not hard-and-fast. In some situations, using costs and benefits to weigh alternatives may be easier to justify than others. For example, if the manager requests a cost-benefit analysis, the choice is an easy one, or if the alternatives have effects that can be handily expressed in monetary units or other equivalent units, using cost-benefit analysis would be appropriate. The most important consideration, however, in undertaking a cost-benefit analysis is that the analyst has an opportunity to provide useful information or useless information. The creativity and care with which the information is gathered, synthesized, documented, and presented can make an enormous difference in the worth of the final product.

Case Studies

Instructions: As Director of Program Analysis at the Highway Traffic Safety Administration, you know how controversial the 55 mph speed limit has been. Your boss must meet with members of state legislatures and the press to discuss the law. He has asked your office to prepare an economic analysis of the law to determine its effectiveness. Working independently, two teams in your office have come up with two studies. You still have time to revise these and synthesize the best of both. Before going home tonight, you and your colleagues sit down to evaluate the studies using the following tasks as guides:

1. Identify the potential use(s) of such an analysis, for example, at which stage of the evaluation process would it be most helpful?
2. Identify and comment on the assumptions made by the analyses.
3. Determine whether or not any decision criteria were made explicit.
4. Identify the benefits and costs in the analyses. Comment on the rationale for them.
5. Assess the level of documentation in the analysis as presented and decide whether or not this would satisfy you if you were the decision maker.
6. Compare these points for each case, identifying any key points of divergence.
7. Evaluate the analyses overall. Which one is better? What else did you want to know about the program?

A COST-BENEFIT ANALYSIS OF THE NATIONAL 55 mph SPEED LIMIT*

The National Maximum Speed Law (NMSL) of 1974, setting the speed limit at 55 miles per hour, had an immediate impact on vehicular speeds. In one year the percentage of vehicles exceeding 65 miles per hour on rural free-flow interstate highways fell from 50 percent to 9 percent. The average speed on those highways dropped from 64 miles per hour in 1973 to 57.6 in 1974. Although compliance with the law was far from complete, we assume that the reduction in speeds was due wholly to the NMSL rather than to the price of gasoline. This report will provide an economic analysis of the costs and bene-

*Adapted from Charles T. Clotfelter and John C. Hahn, "Assessing the National, 55 m.p.h. Speed Limit," *Policy Sciences* 9 (1978): 281–294.

fits of the NMSL. Our cost-benefit analysis is based on actual costs and benefits for 1973 and 1974, rather than on potential benefits if everyone obeyed the law.

COSTS

The major cost of the NMSL is the value of the additional time spent driving as the result of slower speeds. If the law affected only the speed of vehicles, the time cost would be equal to the additional hours spent driving multiplied by the average value of time. But the NMSL law caused some people to cancel trips and others to find alternative modes of transportation, and these options also involve costs. As a result, to calculate time cost based on 1973 mileage would overstate the cost of the NMSL, while cost based on 1974 mileage would be an underestimate. These two figures, however, will permit us to place an upper and lower bound on the estimated economic costs associated with additional driving time.

Using the following formula, where VM is vehicle miles, S is average speed, R is the average occupancy rate per vehicle, and H is the number of hours lost,

$$H = ((VM_{73}/S_{74}) - (VM_{73}/S_{73}))\,R$$

we estimate that the maximum number of hours lost driving in 1974 is 1.87 billion and that the minimum number of hours lost is 1.72 billion.

To estimate the value of time lost in travel, we begin with the average wage rate for all members of the labor force in 1974, \$5.05 per hour. The value of one hour's travel is not \$5.05 per hour because very few persons would pay this sum to avoid an hour of travel. Two studies of the value of travel time estimate that people will pay up to 42 percent and 33 percent of their average hourly wage rate to avoid an hour of commuting. The value of time spent traveling is therefore between \$1.68 and \$2.12 per hour.

Application of the lower cost figure (\$1.68) to the lower time lost figure (1.72 billion hours) results in an estimated travel cost of \$2.89 billion. Using the higher value for time (\$2.12) and the higher hour figure (1.87 billion hours) generates an estimated cost of \$3.96 billion. These are the upper and lower limits of the time cost associated with slower speeds.

Several caveats are in order. We assumed that the value of time is the same for all persons whether they are drivers or passengers, adults or minors. Second, the time costs were based on studies of commuters; time costs for vacationers and long-distance travel may be different. Third, the time costs for commercial vehicles were assumed to be the same as that for commuters.

The NMSL also has some enforcement costs. New signs were posted, people needed to be told about the law, and additional enforcement was required. The latter is especially important because the NMSL is widely violated. Cost estimates from twenty-five states for modification of speed limit signs totaled \$707,000; for fifty states, this results in an estimated \$1.23 million. Spread out over the three-year life of traffic signs, we get an estimate of \$410,000.

The federal government engaged in an advertising campaign encouraging compliance. This cost must be included in the costs of the NMSL. The Federal Highway Administration's advertising budget for 1974 was $2.0 million. We estimate that 10 percent of this, or $200,000, was spent to encourage compliance with the NMSL. We assume that an additional amount of public service advertising time was donated, for a total of $400,000.

Compliance costs are difficult to estimate. The cost of highway patrols cannot be used because these persons were patrolling highways before the NMSL. We assume that states did not hire additional personnel solely for enforcement of the NMSL. As a result, we assume that enforcement of the NMSL will not entail any additional costs above enforcement of previous speed limits.

Our total enforcement costs from patrolling, signs, and advertising, therefore, are $810,000 for the entire United States.

BENEFITS

The most apparent benefit of the NMSL is the amount of gasoline saved. The average gasoline economy improves from 13.3 at 70 miles per hour to 16.9 at 50 miles per hour. We can use this information to estimate the number of gallons of gasoline saved by traveling at lower speeds. Again, we will get a lower bound for gas saved using 1974 figures for miles traveled and an upper bound for 1973 figures. Gallons saved will be calculated by the following formula, where *VM* is vehicle miles traveled and *MPG* is the miles per gallon for the average vehicle speed:

$$G = (VM_{73}/MPG_{73}) - (VM_{73}/MPG_{74})$$

Using the following figures,

$$\begin{aligned} VM_{73} &= 697 \text{ billion} \\ VM_{74} &= 677 \text{ billion} \\ MPG_{73} &= 14.9 \\ MPG_{74} &= 16.1 \end{aligned}$$

the upper limit is

$$\begin{aligned} G &= (697\text{b}/14.9) - (697/16.1) \\ G &= 3.487 \text{ billion gallons} \end{aligned}$$

and the lower limit is

$$\begin{aligned} G &= (677/14.9) - (677/16.1) \\ G &= 3.387 \text{ billion gallons.} \end{aligned}$$

In 1974, the average price of gasoline was 52.8 cents per gallon. This market price, however, does not reflect the social cost of gasoline due to government price controls on "old" domestic oil. The marginal cost of crude oil is the price of foreign oil, not the cost of foreign and domestic oil mixed. Therefore, the price of gasoline must reflect the higher cost of foreign oil. We will use the market price of gasoline in the absence of price controls and entitlements. Our calculations based on the price of foreign crude oil in 1974, refining costs, marketing costs, and retail markup was 71.8 cents per gallon. This figure yields an upper estimate of $2.50 billion and a lower estimate of $2.43 billion.

A major second-order benefit of the 55-miles-per-hour limit was a large drop in traffic fatalities, from 55,087 in 1973 to 46,049 in 1974. Although some of this drop resulted from a reduction in travel, increased daytime driving, and safety improvements, part of the gain must be attributable to reduction in traffic speeds. Studies by the National Safety Council estimate that between 46 percent and 59 percent of the decline in fatalities were the result of the speed limit. Applying these proportions to the decline in fatalities provides an estimate of between 4,157 and 5,332 lives saved.

The social cost of a fatality includes loss of earnings, pain and suffering, loss of home, and loss of family duties performed. Economists have devised ways of estimating the value of a human life. The consensus of several studies is that a traffic fatality costs $200,000 in 1972 dollars or $240,000 in 1974 dollars. Using this figure, the value of lives saved in 1974 was estimated at between $997.7 million and $1,279.7 million.

The NMSL also resulted in a reduction of nonfatal injuries. Since there has been no analysis of the proportion of injury reduction resulting from the speed change, we will use the 59 percent and 46 percent figures found in the fatality studies. Between 1973 and 1974, nonfatal traffic injuries declined by 182,626. Applying the estimated percentages results in a lower limit of 84,008 and an upper limit of 107,749 injuries. Injuries vary in severity and thus in social cost. Generally, three levels of injuries are identified: (1) permanent total disability, (2) permanent partial disability and permanent disfigurement, and (3) nonpermanent injury. In 1971, the proportion of traffic injuries that accounted for injuries in each category was 0.2 percent, 6.5 percent, and 93.3 percent, respectively. The National Highway Traffic Safety Administration estimated that in 1971 the average cost of each type of injury was $260,300, $67,100, and $2,465 respectively. The average injury, therefore, cost $7,182 in 1971 dollars or $8,745 in 1974 dollars. Applying this figure to our injury estimates results in a lower limit of $734.6 million and an upper limit of $942.3 million as the social benefits of injury reduction.

The final benefit of the reduction in vehicle speeds was a reduction in property damage. Between 1973 and 1974, the number of accidents involving property damage fell from 25.8 million to 23.1 million. Because there is no data relating this decline to reduced speeds, we estimated that a lower limit of 25 percent and an upper limit of 50 percent of this reduction was the result of lower speeds. The NMSL saved between 0.65 million and 1.3 million cases of property damage at an average cost of $363 ($300 in 1972 dollars). Therefore, the total benefits from property damage prevented is between $236 million and $472 million.

CONCLUSION

Our estimates of the costs and benefits of the National Maximum Speed Law resulted in the following figures (in millions):

	Lower Limit	*Upper Limit*
Costs		
Time spent traveling	$2,890.0	$3,960.0
Enforcement	0.8	0.8
Total cost	$2,890.8	$3,960.8
Benefits		
Gasoline saved	$2,430.0	$2,500.0
Lives saved	997.7	1,297.7
Injuries prevented	734.6	942.3
Property damage	236.0	472.0
	$4,398.3	$5,212.0

Based on the above figures, we conclude that the benefits of the NMSL exceed the costs. Even using the highest possible cost figure ($3.96 billion) and the lowest possible benefits figure ($4.4 billion) reveals a minimum net benefit of over $100 million. The maximum net benefit of the NMSL is $2.3 billion. The National Maximum Speed Law is cost effective.

*A SECOND COST-BENEFIT ANALYSIS OF THE NATIONAL 55 mph SPEED LIMIT**

The National Maximum Speed Law (NMSL) of 1974, setting the speed limit at 55 miles per hour for all highways, had an immediate impact on vehicular speeds. The average speed on rural interstate highways dropped from 65 miles per hour to under 58 miles per hour within a year. The question is, did the enactment of the law provide more benefits to the nation than the costs it imposed? We believe it did not. We will examine the benefits of gasoline saved, fatalities prevented, and accidents averted. Costs considered include cost of travel time and enforcement. The costs and benefits will be figured for the NMSL's actual impact in 1974.

---

*Adapted from Charles A. Lave, "The Costs of Going 55," *Car and Driver,* May 1978, p. 12.

COSTS

The major cost of the NMSL is the value of the additional time spent driving as the result of slower speeds. Not only did the law affect the speed of vehicles, it also caused some people to cancel trips or find alternative means of transportation. As a result, the social cost of slower speeds may be greater than the value of the additional driving time. Since the speed limit in 1973 was not a restriction, time costs based on actual 1973 miles driven plus 4 percent will provide an upper estimate of time costs (4 percent was the average growth in travel before 1973). A lower limit of the costs of extra travel time will be based on actual miles driven in 1974 (thus omitting costs of trips foregone and alternative transportation).

Using the following formula, where *VM* is vehicle miles, *S* is the average speed, *R* is the average occupancy rate per vehicle, and *H* is the number of hours lost,

$$H = ((VM_{73}/S_{74}) - (VM_{73}/S_{73}))\,R$$

we estimate that the maximum number of hours lost in 1974 is 1.95 billion and that the minimum number of hours lost is 1.72 billion. To find the value of the time lost in travel, we will use the average wage rate for all members of the labor force in 1974, $5.05 per hour. We will not take a percentage of this figure based on what commuters would pay to avoid an hour of travel (the normal figure is 42 percent of the hourly wage rate). We avoid discounting the value of people's time for two reasons. First, the value of time in cost to society is equal to what society will pay for productive use of that time. Time's value is not what a commuter will pay to avoid commuting because commuting has other benefits, such as solitude for thinking or the advantages of suburban living. Second, the value of time spent driving for a trucker is many times the industrial wage rate. Discounting would greatly underestimate the value of commercial drivers.

Applying the value of one hour's time to the lower limit of hours lost (1.72 billion hours) results in an estimate of $8.686 billion. Using the upper limit for time lost (1.95 billion hours) results in the upper and lower limits of costs of increased driving time.

The NMSL also has some enforcement costs. New speed limit signs were posted, people were urged to comply with the law, and additional enforcement was required. Although the enforcement and compliance efforts met with limited success, making the NMSL the most violated law since Prohibition, these costs must be included as costs of the NMSL. Cost estimates from twenty-five states for the modification of speed limit signs totaled $707,000. Assuming that the other twenty-five states incurred similar costs, a total estimate is $1.23 million. This cost cannot be amortized over the life of the signs since none of these signs would have had to be replaced if the NMSL had not been enacted. Replacement costs in the following years are assumed to be normal.

The federal government engaged in an advertising campaign encouraging compliance. "It's not just a good idea, it's the law," "A law we can live with," and "Saves lives, money, energy" were slogans heard by all Americans. The Federal Highway Administration's advertising budget for 1974 was $2.0 million. Because this agency does little other advertising, we assumed that all of this was spent to encourage compliance with the new speed limit. We also assume that an additional amount of public service advertising time was donated by broadcasters for a total cost of $4 million.

Compliance costs pose some problems, but they can be estimated. In 1973, some 5,711,617 traffic citations for speeding were issued in the United States. In 1974, this number of citations jumped by 1,713,636 to over 7.4 million. Each additional traffic citation includes an opportunity cost to society. If a law enforcement officer were not issuing traffic tickets, he could be solving other crimes. Assuming that it requires 15 minutes for a law enforcement officer to issue a speeding ticket, the total cost of extra enforcement is $2.9 million. This figure is based on the average cost of placing a law enforcement officer on the streets at $6.75 per hour. This figure is clearly an underestimate because it does not count time lost waiting to catch speeders.

Approximately 10 percent of all speeders will demand a court hearing. Estimating an average of thirty minutes for each hearing and hourly court costs of $45 results in an additional cost to society of $3.8 million for 171,000 cases. Given the overloaded court dockets, this opportunity cost may be even higher.

Total enforcement and compliance costs for the NMSL are estimated to be $12 million. This includes extra enforcement, advertising, and sign changes.

BENEFITS

The intended benefit of the NMSL was a saving of gasoline. We will *not* estimate gasoline saved by comparing 1973 and 1974 miles-per-gallon figures in relation to vehicle miles traveled. The federal figures for average miles per hour are estimates based on several assumptions. Given the conflict between industry estimates, Environmental Protection Agency estimates, and Energy Department estimates, any miles-per-hour estimate must be considered unreliable. The number of vehicle miles traveled is also based on gallons of fuel sold multiplied by average miles per hour. Hence, this figure is also subject to error.

Donald Rapp of the University of Texas at Dallas has analyzed the efficiency of gasoline engines. He concluded that the effect of reducing the average speed on free-flow interstate highways (and corresponding reductions elsewhere) would save 2.57 percent of the normal gasoline used. In 1974, American motorists consumed 106,301 million gallons of gasoline. Saving 2.57 percent would total 2,732 million gallons.

In 1974, the average market price for a gallon of gasoline was 52.8 cents. This implies a saving of $1,442 million. We use the market price of gasoline rather than the price based on foreign crude oil because there is no way to determine whether a marginal gallon of gasoline will be imported or come from domestic reserves. In addition, the costs and benefits of the NMSL should not be distorted simply because the U.S. government does not have a market-oriented energy policy. In 1974, gasoline cost 52.8 cents per gallon, and therefore a gallon of gasoline saved was worth 52.8 cents.

A second benefit of the new speed limit was a large drop in the number of traffic fatalities from 55,087 in 1973 to 46,049 in 1974. How many of this 9,038 drop can be attributed to the new speed limit is not known. Some analysts have argued that all of this 9,038 drop can be explained by other factors, such as safety improvements, less travel, and weather. Others argue that the NMSL saved all 9,038 lives. We will take a middle position and accept the National Safety Council's estimate that between 59 percent and 46 percent of the decline was the result of the new law. Therefore, the lower estimate of lives saved is 4,157 and the upper estimate is 5,332.

Assigning a value to a life saved requires estimating such things as the loss of earnings, pain and suffering, and loss of companionship. The consensus of studies by economists, including a study endorsed by the Department of Transportation, is that a traffic fatality cost $240,000 in 1974 dollars. Using this figure, the benefits of saving lives for 1974 were between 997.7 million and $1,279.7 million.

The NMSL also resulted in a reduction of nonfatal injuries. We will use the National Safety Council's percentages for reductions in deaths and attribute 46 percent to 59 percent of the reduction in nonfatal injuries to the NMSL. Using a total reduction of injuries of 182,626, the estimated number of injuries prevented was between 84,008 and 107,749.

Injuries vary in severity and therefore in social cost. Generally, three levels of injury are identified: (1) permanent total disability, (2) permanent partial disability or permanent disfigurement, and (3) no permanent injury. In 1971, the proportion of traffic injuries accounted for by each of these three types was 0.2 percent, 6.5 percent, and 93.3 percent, respectively. According to the National Highway Traffic Safety Administration, the average cost in 1974 dollars of each type of injury was $316,948, $81,702, and $3,001, respectively. One problem with these estimates is that a permanent total disability incurs greater social costs than a fatality. From a cost perspective, with these figures it would be efficient to sponsor euthanasia programs for all permanently disabled. To avoid this inconsistency, the cost of a total permanent disability was reduced to $240,000. The average cost of a traffic injury was computed to be $8,591. The social benefit of reduced injuries therefore had an upper limit of $926 million and a lower limit of $722 million.

The final benefit of the new law was a reduction in property damage. Between 1973 and 1974 the number of accidents involving property damage fell from 25.8 million to 23.1 million. No study has related this decline to a decline in traffic speeds. Since most of the property damage claims result from collisions at low speeds (the average cost is only $363 in 1974 dollars), we will generously assume that 25 percent of the reduction could be attributed to the slower speeds. The total benefits of avoiding 0.65 million accidents is therefore $236 million.

CONCLUSION

Our estimates of the costs and benefits of the National Maximum Speed Law resulted in the following figures (in millions):

	Lower Limit	*Upper Limit*
Costs		
Extra travel time	$8,686.0	$9,848.0
Enforcement costs	12.0	12.0
Total cost	$8,698.0	$9,860.0
Benefits		
Gasoline saved	$1,442.0	$1,442.0
Lives saved	998.0	1,280.0
Injuries prevented	722.0	926.0
Property damage	236.0	236.0
	$3,398.0	$3,884.0

Based on the above figures, we conclude that the benefits of the NMSL do not exceed the costs. Even with the highest possible benefits ($3.9 billion) and the lowest possible costs ($8.7 billion), the NMSL has costs twice as great as its benefits. Costs exceed benefits by between $4.8 billion and $6.5 billion. The National Maximum Speed Law is *not* cost effective.

FOR FURTHER READING

Kellman, Stephen. "Cost-Benefit Analysis: An Ethical Critique." *Regulation* 5 (January–February 1981): 33–44.

Little, I. M. D. *A critique of welfare economics.* London: Oxford University Press, 1957.

Meier, Kenneth J. "The Limits of Cost-Benefit Analysis." In *Decision making for public administrators,* edited by Lloyd G. Nigro. New York: Marcel Dekker, 1984.

Mishan, E. J. *Cost-benefit analysis.* New York: Praeger, 1976.

Pigou, A. G. *The economics of welfare.* London: Macmillan, 1932.

Rhoads, Steven E. *Valuing life: Public policy dilemmas.* Boulder, Colo.: Westview Press, 1981.

Stokey, Edith, and Zeckhauser, Richard. *A primer for policy analysis.* New York: Norton, 1978.

Thompson, Mark S. *Benefit-cost analysis for program evaluation.* Beverly Hills, Calif.: Sage Publications, 1982.

Chapter 4

EVALUATION DESIGNS

This chapter begins with a discussion of internal and external design validity as these concepts are related to the needs of the evaluator, program staffs, and external consumers. We then turn to the various threats to achieving a valid design. Next, various designs are presented, including experimental designs, followed by nonexperimental approaches such as time series, nonequivalent control groups, and matched-pair designs. The chapter concludes with the participant-observer approach and designs that mix outcome and process measures.

VALIDITY

The subject of design validity can be divided into the two broad categories: internal validity and external validity. Internal validity asks whether the evaluation measures what is intended. External validity asks whether the findings can be generalized, that is, whether the results of the evaluation can be used to make judgments about other programs and whether other researchers can replicate the methodology and the findings.

INTERNAL VALIDITY

Internal validity should be the principal concern of the researcher. Social scientists generally seek to ensure that their research designs actually measure what is intended and not some extraneous phenomenon; moreover, they want to employ the most powerful available measure to ensure the quality of their findings. In evaluation research, however, internal validity involves the additional concerns of correctly identifying program goals and developing a design that meets the needs of the program staff. Internally valid designs should also produce findings that are comprehensible to the consumers of the evaluation.

Correct identification of program goals can best be accomplished in consultation with program staff. Because program goals are often vague, missing, or contradictory, failure to engage in goal consultation may produce a methodologically valid research design that measures phenomena peripheral to the program.

Evaluation specialists must also answer the research questions in a form that is understandable and acceptable to the consumers of the evaluation. Consumers include political executives and legislators as well as program staffs. Because these consumers are not evaluation experts, findings must be expressed in clear, precise language. In addition, external consumers may have program goals that are different from the goals of program administrators. Designing an evaluation that is acceptable to all audiences should, therefore, be of critical concern to the evaluation specialist. Failure to take into account all the interests may doom an evaluation to controversy or, worse, to being ignored from the outset.

EXTERNAL VALIDITY

External validity involves the interdependent concerns of replication and inference. Replication is of particular concern to research purists because the implicit consumers of the findings are their research peers who may wish to replicate the research. To assure replication of the methodology and findings, the design and conduct of the research must be documented systematically. This precision makes it possible to have repeated replications that may result in a body of data sufficient to build a theory regarding the phenomenon in question. Evaluations are replicated to ensure that the findings are not unique to a particular time or place.

Another concern of external validity is inference, or whether the findings of the evaluation can be generalized to all similar programs. Inferential or theoretical knowledge is the goal of research in both the physical sciences and the social sciences. Scientific investigation therefore seeks to discover relationships among elements in the phenomenon under investigation. For example, if multiple replications of research findings demonstrate a causal relationship between Element A and Element B (for example, poverty and malnutrition), researchers will infer that, other things being equal, when A is present B will occur.

Unfortunately, evaluation researchers do not yet share the scientist's concerns about replication and inference. If they did, identical research designs would be used when several evaluation teams examined separate delivery units of the same program. Such replication would make possible a meaningful assessment of program success. Many evaluations, however, are idiosyncratic, tailored to a specific program unit.

Without consistent evaluation strategies, decision makers cannot be sure that program officials are pursuing identical or even similar program goals, or that there is any commonality among program structures. The use of accepted and replicable evaluation designs would be very useful to evaluation consumers as well as to evaluation professionals.

THREATS TO VALIDITY

There are a number of specific threats to internal and external validity which researchers can avoid. Certain factors in the formulation and conduct of an evaluation can serve to invalidate the findings. These factors may be elements in the environment over which the researcher has no control, but threats to validity are more likely to stem from the methodologies employed by the researcher.[1]

HISTORY

The term *history* is used to describe validity problems that result when the program's environment changes after the program is introduced. When external changes occur simultaneously with the program, the causal impact of the program cannot be separated from the environment. These environmental factors may distort the apparent impact of the program and lead the researcher to believe that the program is a success or failure when in fact the researcher is measuring two different phenomena.

Suppose that one wished to evaluate an experimental program designed to assist persons on welfare to become work ready, to assist them in finding suitable employment, and to help them with such logistical problems as child care and transportation. In order to measure the success of the experimental program, one might begin by gathering baseline data on the employment success rates of welfare recipients who had sought work under previous programs. If during the experimental program the economy of the region experienced a significant downturn, however, persons with work experience who might other-

[1] The discussion of threats to validity relies heavily on the taxonomy of threats developed by Donald T. Campbell and Julian C. Stanley in *Experimental and quasi-experimental designs for research* (Chicago: Rand McNally, 1963).

wise have been employed would be competing for jobs with welfare recipients who participated in the experimental program. Data on program participants might not compare favorably with data gathered on welfare recipients before the economy faltered.

In this example, history is the economic downturn. History could have been controlled for by gathering data on nonparticipants at the end of the experimental program as well as at the beginning. Program participants may well have achieved more job placements than other welfare participants during the downturn.

MATURATION

Maturation is another potential threat to validity. The term *maturation* reflects the passage of time and concomitant changes in the subjects under study. Controlling for maturation is particularly important in research projects involving children because children grow stronger and taller and more capable of reasoning and learning with the passage of time. For example, childrens' vocabularies and their cognitive development are affected by age. Simple before and after tests of subjects in experimental reading programs, therefore, would measure the effects of maturation as well as the effects of the program. Comparisons should therefore be made between children receiving the program and their age cohorts who do not receive it. Merely gathering before-and-after data on the experimental children will not yield valid results.

Maturation also can affect adult subjects insofar as it refers to development over time. This development need not be positive. In the sense that it is used here, maturation could reflect negative changes, as when an adult becomes more hungry or grows more cynical over time. Using the job placement program as an example, the inability of participants to find work may cause them to assess the skills they obtained in the program negatively. The cynics might give up the search for work, causing the program to fail not as a result of program content but as a result of cynicism (maturation).

TESTING

Pretesting subjects can affect their performances on subsequent tests. This factor is particularly important in testing children, because the pretest may be a learning experience for them. A math skills test, for example, allows children to practice their math and possibly improve their performance on the posttest as a function of the pretest learning experience. Among adult subjects, testing may alert them to the fact that something is going to change. Subjects become sensitized to the variables being studied. This phenomenon, sometimes known as the Hawthorne effect,[2] is likely to occur in testing experimental management or personnel programs. The act of testing may lead to a change in the climate of the workplace, which in turn may alter work outcomes. Before-and-after testing on the subjects can have an impact independent of the effects of the experiment. A carefully constructed design can control for the impact that testing can have on the experiment (see the discussion of experimental designs later in this chapter).

INSTRUMENTATION

Instrumentation is the phenomenon in which the calibration of the measuring instrument is altered. Variations in findings may also result from turnovers in the research staff or even changes within the person doing the evaluation.

A change in instrumentation may be mandated by a change in public policy. In the 1970s, for example, state rehabilitation agencies had to change their accounting procedures as a result of a federal requirement that 50 percent of all rehabilitation clients be severely handicapped. Previously, the number of placements was the only criterion, regardless of the severity of the disability. Providing expanded services to the severely handicapped required changes in program delivery and in the means used to measure success. There were several possible measurement alterations. Preparing the severely handicapped to maintain independent residences might supplement the job placement criterion as a measure of agency success, or a hard-to-place handicapped person might be counted double in measuring program outputs. On the other hand, simple comparisons between programs using the new and the old measures would result in a dramatic increase in the apparent performance of an agency using the new measures. This increase would occur even though overall agency performance remained relatively unchanged.

A similar change in instrumentation hindered the evaluation of the 1974 National Maximum Speed Limit (the 55-miles-per-hour limit). Before the law, a traffic fatality was considered a death within one year after the accident. After the law, a thirty-day limit was used. The change in instrumentation hindered evaluation of the law's impact on traffic fatalities.

Turnover in evaluation staff is another type of instrumentation change that can be dealt with by carefully orienting incoming staff regarding goals of the

---

[2] The Hawthorne effect is so named as a result of studies conducted on workers at the Hawthorne Works of Western Electric Company during the 1920s and 1930s. The Hawthorne research team concluded that some of the work behavior of the subjects was a result of their knowing they were being treated differently from normal. The findings of the Hawthorne experiments were reported by F. J. Roethlisberger and W. J. Dickson in *Management and the worker* (Cambridge, Mass.: Harvard University Press, 1939).

research and the rationale for the measuring instruments used. New staff must also be apprised of the proper methods for applying the research instrument.

Prolonged contact with the program and its staff may cause changes within the same rater: An objective evaluator may become a program advocate. This problem is particularly acute when the evaluator's personal assessment of subject performance is the measure of program success.[3] Such problems can be corrected through careful training and reliability checks on the raters.

STATISTICAL REGRESSION

Statistical regression threatens validity when extreme performers are the experimental subjects. On any given test, individual performances will vary as a matter of luck. For example, on a good day a superior student will guess correctly on the answers he or she does not know, and on a bad day the opposite will occur. The same is true at the other end of the continuum: Those scoring the lowest on the exam will do so partly as a result of inability and partly as a result of bad luck.

Suppose that a researcher selected the poorest performers on a reading examination for inclusion in a special tutorial program after which the students were to be retested. Because random changes would cause some of the students to improve their performances on the second test regardless of the merits of the program, a before-and-after test of the poorest performers would not show which students' performances could be accountable to luck and which were a result of program participation.

Consultations with program staffs can help avoid the problems of statistical regression. Program staffs may believe that the targets of the program should be the most needy, so the evaluation team must explain that the best way to determine program success is to select students of varying abilities on a random basis for inclusion in the experimental and control groups.

SELECTION BIAS

Selection bias refers to the danger of making comparisons between groups that are erroneously believed to be equivalent. Suppose, for example, that the police department in City A instituted a new crime detection training program and that one year later the arrest and conviction rates of the officers working for

---

[3] This phenomenon, also known as "going native," is precisely why some researchers argue that outcome evaluations which use objective measures are the most valid approach to evaluation. In addition, virtually every evaluation text argues for the importance of maintaining the evaluator's detachment and professional objectivity.

City A were compared with those of City B. This after-only design could not guarantee that the differences in performance were a function of the training program. Officers in City A may have been better at crime detection than officers in City B before the training program was instituted.

Selection bias is a possible problem whenever participants in an experimental program are volunteers. For example, drug addicts who volunteer for an experimental behavior modification program to reduce drug dependency cannot be compared with another randomly selected group of addicts. The volunteers are more motivated to do something about their addiction, which distinguishes them from other addicts. A more appropriate experiment would be to assign volunteers randomly to the experimental program and to a placebo control group (of course, such a study could be generalized to volunteer groups only, because nonmotivated addicts would resist participation).

EXPERIMENTAL MORTALITY

Experimental mortality is one of the most vexing problems for evaluation researchers engaged in comparison of pilot programs. "Mortality" in this case refers to subjects' dropping out of the program. Returning to the job-training example, persons with low-level motivation might drop out of an experimental program at higher rates than persons in ongoing programs simply because the pilot program requires an extra effort from the subjects. The threat to validity is that subjects who remain in the pilot program are more motivated than subjects in the comparison group. The probability that this group will find jobs at a higher rate is therefore increased as a function of the dropout of the unmotivated rather than as an effect of the program.

An additional problem can result from the good intentions of program staff who may add subjects to the experimental group after the experiment is under way. The additions would not receive the same length and type of training and might therefore pull down the performance of the experimental group. The staff might also add subjects out of a desire to extend program benefits to the largest possible group. The humanistic impulses of the staff might also incline them to extend program benefits to their most needy clients, which would add the problem of selection bias to that of mortality. Whatever the motivation, the principal effect of subject additions or substitutions is contamination of the research findings, occurring because the staff is ignorant of the need to preserve the purity of experimental conditions. Close cooperation between program and evaluation staffs throughout each phase of the experiment can help avoid such problems. After the experiment runs its course, the staff can extend the program to all clients, but substitutions before that time will never permit the manager to conclude that the program had the desired effect.

INTERACTION OF SELECTION WITH OTHER VARIABLES

Selection can interact with other variables to compound the problems of validity if selection is based on factors other than random assignment. For example, a test to determine the impacts of a teaching method on college sophomores, juniors, and seniors suffers from several potential interaction effects. The most obvious is the maturation variable. Juniors and seniors are not equivalent to each other, or to sophomores. Each successive year in school increases the probability that a student will become a serious student and develop quality study habits. The ranks of the lower classes, moreover, are populated by a greater number of persons who refuse to take school seriously and who will not finish. The program evaluator should examine any nonrandom selection process and ask if selection interacts with mortality, regression, instrumentation, testing, maturation, or history. If so, the study must be redesigned or the conclusions qualified.

FOUR THREATS TO EXTERNAL VALIDITY

Donald Campbell and Julian Stanley discuss four threats to the external validity of research. External concerns are important when the research findings are to go beyond the specific setting of the program. With pilot programs, however, it is assumed that the findings will be manifested in large expenditures for wider applications of successful programs. For this reason alone, the researcher must provide for as much external validity as possible.

INTERACTIVE EFFECTS OF SELECTION BIAS AND THE EXPERIMENTAL VARIABLE

A major threat to external validity is the interactive effects of the experimental variable and the selection bias. To make sure that such effects do not occur, one must use a selection process based on random assignment to the experimental conditions from a population sufficiently large to ensure representativeness. For example, testing the benefits of Nautilus weight-training equipment for football players using the University of Oklahoma football team as the experimental group would jeopardize the generalizability of the results. Members of the Oklahoma team are usually dedicated athletes attending college on scholarships. A sizable portion of them also aspire to careers as professional athletes. The team is probably more motivated than would be the case with college teams generally.

The interactive effects of the experimental variable and the selection can also be illustrated with the job-training example. Suppose that the research team sought to control as many conditions as possible to isolate the effects of the program and that therefore the evaluation team requested volunteers from existing jobs programs. And suppose that the evaluation team then divided the volunteers into an experimental group and a control group, the experimental group receiving the special training and the control group continuing as they were. Selection bias would have occurred because both the experimental group and the control group were taken from persons sufficiently motivated to participate in existing jobs programs and to volunteer to be part of the experimental program. The study is apt to show that the experimental program produced results not much greater than current programs because, while the experimental group was receiving job readiness training, the equally motivated control subjects would be looking for work.

Ideally, both experimental subjects and control subjects should be selected at random from among the general welfare population, but human subjects cannot be compelled to participate. In fact, there are numerous federal restrictions on the use of human subjects in program experimentation. Nevertheless, a valid evaluation of a program could be achieved by comparing the success rates of the experimental group with those of the control group volunteers, and the success of both groups with that of the general welfare population.

REACTIVE OR INTERACTIVE EFFECTS OF PRETESTING

Pretesting subjects may make them either more susceptible or less susceptible to the effects of the experimental condition. For example, a pretest of employee attitudes regarding supervisor approachability and effectiveness might start subjects thinking in those terms when they had not done so in the past. Such heightened awareness would sensitize employees to changes in managerial practices whereby supervisors sought to be more approachable and supportive. The subjects might therefore rate a supervisor higher on these traits than they would have had there not been a pretest.

REACTIVE EFFECTS OF THE EXPERIMENTAL ENVIRONMENT

Reactive effects of the experimental environment that cannot be duplicated in the nonexperimental condition can also invalidate research findings. For example, the presence of the evaluation team may cause the subjects to deviate from normal behavior, and subjects who know that they will be posttested may put forth extra effort.

The experimental environment also may be so controlled that it does not reflect the real world. For example, much of our knowledge about human performance and psychological motivation has been gained in laboratories using homogeneous subjects. Persons seeking to apply this knowledge of human performance in groups are frequently frustrated because tasks, reward systems, and accountability operate differently under experimental conditions and real-world conditions. Experimental groups composed of student volunteers may be more cooperative or be willing to take greater risks than persons in real-world groups composed of supervisors, subordinates, and peers. In the real world, moreover, leadership will most likely be asserted by the group's supervisor. And group decisions in the real world are influenced by relationships among group members that existed before the group assumed the current task. Risk taking and the degree of cooperation in the real world may have been defined and limited by the organization. Given the above, program innovations that prove promising under experimental conditions should be pilot tested under real-world conditions before being broadly applied.

MULTIPLE-TREATMENT INTERFERENCE

Multiple-treatment interference is another threat to validity. Because the application of consecutive treatments is cumulative in the subjects, the researcher frequently has difficulty isolating the effects of subsequent treatments from earlier applications.

For example, suppose that a manager wished to know if employees are more productive when they are compensated on an hourly basis as individuals, when they are compensated on a piecework rate on the basis of their individual productivity, or when they are compensated as a unit on the basis of group productivity. And suppose that the manager decided to experiment by paying workers in a particular unit by the hour for one week, and then individually by the piece for one week, and then as a group for one week, measuring at the end of each week both productivity and job satisfaction. The data on employee productivity and satisfaction when paid by the piece as individuals would be tainted by the employees' experience of being paid by the hour as individuals. Similarly, employee attitudes toward compensation by the piece as a group, and their productivity, would be colored by the previous two experiments. The most that could be learned from the experiment is how groups react when subjected to the three treatments in the same order. The pitfalls of multiple-treatment interference could have been avoided by conducting three separate studies of workers who were randomly assigned to the three conditions.

Multiple treatments are not always multiple treatments by the program manager. Sometimes subjects receive another treatment by an outsider. For example, the federal Office of Personnel Management (OPM) wanted to know the effect of the Senior Executive Service (SES) system on the morale and work effectiveness of senior federal managers. The SES was created by the 1978 Civil Service Reform Act. The SES system was to combine a system of increased policy formulation responsibility with a system of merit pay and bonuses and, hopefully, increase productivity and accountability. Independently of the OPM, Congress limited bonuses paid to higher-level managers. The civil servants were therefore subjected to two treatments: the Senior Executive System and the bonus cap. Only one was intended.

DESIGN SELECTION

In addition to being concerned with internal and external validity, the evaluator also must make the design programmatically valid. Program validity occurs when the researcher generates an evaluation strategy that rules out as many validity pitfalls as possible while producing data that is both understandable and useful to the program staff. Evaluation costs must also be within the resources of the program budget. Selecting a design that is rigorous enough to avoid validity threats is the prime concern of evaluation specialists. Making the findings relevant to the consumers, however, requires cooperative effort in the design phase between evaluators and program staff. The design thus can reflect the needs of program administrators as well as evaluator concerns for design validity.

In this section, several evaluation designs are presented, but there is no attempt to judge the relative merits of the designs. The yardstick by which evaluation designs should be measured is how applicable they are to the needs of the program, not their methodological sophistication.

EXPERIMENTAL DESIGNS

The Classical Experiment. Experimental design is the evaluation method that most closely approximates the rigorous standards for research found in the physical and social sciences. All experimental designs feature random assignment of subjects to experimental and control group conditions. Random assignment is necessary to ensure that history, maturation, selection biases, and the like do not threaten the validity of the data. In addition, the experimental strategy carefully controls the exposure of subjects to the experimental condition.

The classic pretest-posttest experimental design can be illustrated as follows:

R O X O
R O   O

The two Rs represent two groups of subjects who have been randomly assigned to the experimental and control group condition. A pretest is administered to both groups before the experimental condition is introduced. The pretest observation is represented by the Os in the second column of each condition. Only the upper row contains an X, which represents the experimental condition and thus indicates the experimental group. Both the upper and lower rows contain an O in the last column, indicating the posttest measurement.

Suppose that one wished to learn the impact of jogging on pigs that have high-cholesterol diets. The researcher might randomly assign pigs of comparable age, weight, and gender to the experimental and control conditions. In this case, X would be jogging. All pigs would have complete physical exams at the outset, which would constitute the pretest. Pigs with abnormalities would be screened from the test.

The next step is the conduct of the experiment. The experiment requires that both the experimental group and the control group be fed diets high in cholesterol. The groups would be housed separately to ensure the purity of the experimental conditions. (It would not do to have a control subject undergo the rigors of jogging.) The experimental pigs would be taken out jogging each day for increasing distances until the test conditions of jogging fitness were reached. The control group would not jog (presumably some of the pigs might play golf, tennis, or some other sport).

The final test would be to give the experimental and control groups complete physical examinations again to determine the level of cholesterol in the bloodstream and the overall cardiovascular condition of the subjects. The before-and-after condition can be represented as follows:

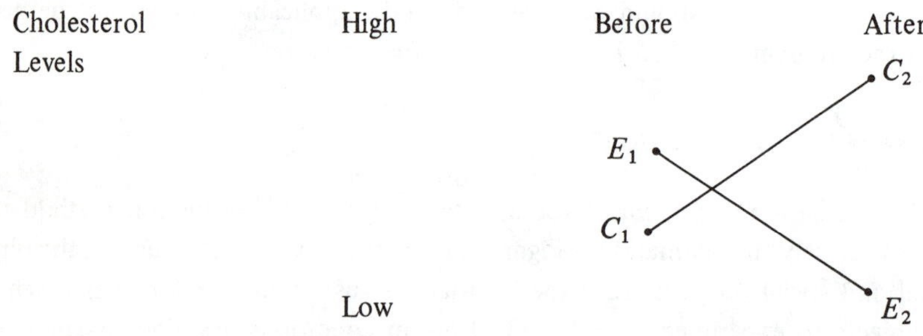

The dots at E_1 and C_1 respectively represent the cholesterol levels in the experimental and control subjects before introduction of the experimental variable of a high-cholesterol diet. E_2 and C_2 represent the posttest cholesterol levels in the two groups. The researcher would measure the relative changes in cholesterol level between the two groups to determine the effects of jogging on high-cholesterol diets. The plot indicated above shows that jogging resulted in reduced cholesterol levels despite the increase of cholesterol in the diet. The control group, which did not jog, experienced increased cholesterol.

The test could be made more complex by introducing additional experimental and control conditions. For example, the researcher might also want to compare the above results with pigs that were fed regular diets but did not jog, pigs that received a normal diet and did jog, and a group that received a low-cholesterol diet but did not jog. The number of variations on the design would be limited primarily by the number of questions the researchers wished to investigate and the number of pigs the research budget could afford.

The Solomon Four-Group Design. With human subjects, the impact of testing becomes an important consideration, especially when the experiment calls for learning and the measurement of learning. To ensure against the interactive effects of testing, Campbell and Stanley offer the Solomon four-group design. This design uses two experimental and two control groups which can be represented as follows:

R	O	X	O
R	O		O
R		X	O
R			O

Assignment to the various conditions is done on a random basis and is indicated by the Rs in the first column. The first group receives a pretest, undergoes the experimental condition, then receives a posttest. Group Two receives only the pretest and the posttest and acts as a control on Group One. Group Three undergoes the experimental condition but receives only a posttest. Group Four receives only the posttest.

The methodology is cumbersome, but it does rule out the interaction effect of testing and X. Differences can be measured between Group Three, which received only the experiment and the posttest, and Group One to determine if X interacted with testing. Group Four provides information on comparable subjects who have undergone the same maturation and historical conditions without the experimental or testing impacts. Comparing Group Two and Group Four would provide a pure measure of the impact of simple testing on posttest performance without the impact of X.

The Posttest-Only Design. A simpler method for ruling out the interactive effects of X and testing is the posttest-only approach. This approach involves random assignment to two groups, the administration of the experimental condition to one group, and the administration of a posttest to both the experimental group and the control group. The posttest-only design is diagrammed as follows:

R	X	O
R		O

Posttest-only designs are preferable to Solomon four-group designs when the researcher wishes to rule out only the impact of testing.

A researcher may design two discrete experiments so that the experimental group in one condition acts as the control group for the other experiment, and vice versa. The mutual control design has been used with success in educational situations.[4] For example, the design can be achieved by randomly assigning students to two groups, A and B. Group A undergoes mathematics training but is not given a pretest. Group B undergoes reading training but does not take a reading pretest. At the end of the first training cycle, Group A is tested for mathematics gain and Group B is tested for reading gain. Each group then becomes the control group for the other by using a pretest-posttest approach in the second round of training. In the second round, Group B begins math training but, unlike members of Group A in the first round, takes a pretest. Group A begins its reading training and takes a pretest. The pretest scores from the second phase become the control group scores for the first round. The design can be represented as follows:

Math	R_A	X	O	Math	R_B	O	X	O
Reading	R_B	X	O	Reading	R_A	O	X	O

Benefits and Problems of Experimental Designs. The benefits of experimental designs are the confidence one gains in the internal validity of the findings and the range of variables that can be tested using additional experimental conditions and control groups. In evaluating social programs, however, the cost and difficulty of sustaining the experimental conditions must be weighed against the relative importance of the knowledge to be gained from the experiment.

Sustaining experimental conditions over a period of time is the principal difficulty in applying experimental designs in evaluation research, especially when the subjects are free to come and go or drop out of the experiment. On the other hand, experimental designs can be especially useful when subjects are not mobile, as in the case of long-term patients in mental institutions or convalescent hospitals or when the subjects are prisoners (in such cases, however, generalizations can only be made to these populations).

Experimentation also may be useful when there is a high possibility of contamination from the history and/or maturation threats to validity. For example, a program of counseling fifteen- to eighteen-year-old gang members might report spectacular success rates if the only measure was continued gang membership after counseling, but an experimental design might indicate that as adolescents

---

[4] See, for example, Patricia Kendall, "Evaluating an Experimental Program in Medical Education," in *Innovation in education*, ed. Matthew B. Miles (New York: Teachers College, 1969), pp. 343–360.

mature they tend to discontinue gang membership without counseling. What was attributed to program success was actually the result of maturation.[5]

Staff concerns for program delivery can also pose a threat to sustaining the experimental condition for the duration of the evaluation. Suppose that an experimental drug treatment program was the subject of evaluation. Outpatients at the drug treatment center would be assigned to the treatment and control groups on a random basis. Suppose further that the period of experimentation was to be two years. If the program seemed to be a success on the basis of an interim report, the staff might press for immediate delivery of the benefits to all patients in the center, including the control group. If program administrators yielded to staff pressure, the evaluator would have no alternative but to terminate the experimental conditions. This problem can be avoided by laying out the need for preserving the experiment at the outset of the design phase and by eliciting a staff commitment not to press for early termination of the experiment.

An Example of Experimental Design. Despite the possible problems with experimental designs, there are circumstances in which one or more pilot programs can be worth conducting using the experimental method or a close approximation of it. The drug rehabilitation program is an example of such a program.

Suppose that one wished to determine whether a halfway house for drug users is more beneficial than outpatient centers, where treatment and counseling normally take place. The first step would be to identify the variables that might affect program outcomes, for example, whether or not the halfway house is to use heroin substitutes, whether to utilize social workers or former addicts as counselors, and whether the halfway house will conduct in-house job training or operate a job placement service.

The next step would be random assignment of subjects to the experimental and control conditions. One could generate a list of potential participants from the existing centers, contact the persons in question, and invite them to participate in the program. Persons who agreed would be assigned to the experimental or control conditions. Control subjects would continue as they were at the regular treatment center, experimental subjects would participate in the halfway house experience.

Two potential dangers to the success and validity of the study have already cropped up: First, it is unlikely that persons who are assigned to the control

---

[5] For a treatment of the problems faced by persons seeking to deal with delinquency, see Walter B. Miller, "The Impact of a 'Total Community' Delinquency Control Project," *Social Problems* 10 (1962): 168–191, or Edwin Powers and Helen Witner, *An experiment in the prevention of delinquency: The Cambridge-Somerville youth study* (New York: Columbia University Press, 1951).

condition will agree to increased scrutiny (for example, psychological tests) when they see no apparent benefit. An alternative would be to select a random sample of participants from among those agreeing to participate and to use the general population of the regular treatment program as a control, but this limits the range of measures that one can use because there is no direct access to the control group.

The second threat to the validity of the outcome is that the volunteers who enter the halfway house must be selected from among those agreeing to participate and therefore are not equivalent to volunteers for medical experiments, where the mental attitude and emotional stability of the volunteers have no effect on their physiological reactions to one treatment or another. On the other hand, volunteers selected from among heroin users may have the motivation to terminate their drug dependency. It is possible that the same persons would have achieved a high success rate at the drug treatment center if they had not been invited into the experimental program.

The research, therefore, cannot be thought of as purely experimental, even though the conditions are more controlled than in some of the nonexperimental designs treated later in this chapter. Given these limitations on experimental purity, it would be best to run several tests of the program simultaneously. If a heroin substitute is to be used in conjunction with the halfway house, a separate group could be given the heroin substitute but receive counseling at the regular treatment center. Similarly, halfway houses without the substitute could also be run and the use of professional and former-addict counselors could be varied. Increasing the variation in conditions would strengthen the findings.

Once the experimental conditions have been designed, the next step is to develop proximate indicators of program success. The number of cases in which the subjects experience complete freedom from drug addiction, if taken alone, would be a stringent test of program success. Other goals of the program could be making addicts productive citizens, reducing crime, and minimizing the return to complete addiction. Additional indicators of program success might include the ability of a subject to hold down a job and the number of arrests of control and experimental group members. A final indicator might be the rates with which subjects abandon various forms of treatment to resume the life of complete dependency on heroin.

The findings of the above design would provide decision makers with information about the efficacy of expensive labor intensive programs before large expenditures of public funds were poured into wider applications of the program. Unfortunately, funding limitations do not always permit experiments that are rigorous enough to isolate all the control conditions. Public officials must therefore be convinced that rigorous experimentation is a necessary decision-aiding tool in the process of forming policy.

NONEXPERIMENTAL DESIGNS

Nonexperimental designs incorporate the systematic data-gathering techniques and rigorous indicators that are found in experimental design, but because they are not constrained by the requirements of random assignment and maintenance of experimental conditions, nonexperimental designs are particularly attractive to program evaluation specialists. Furthermore, evaluators are frequently called on to begin their evaluations after a program has been in operation for a number of years, which makes experimentation impossible. Even some pilot programs are so broad-based in scope (for example, city or statewide) that random assignment is impossible. Fortunately, time series, nonequivalent control groups, matched-pair designs, and the like can provide officials with information about the success of the program without the rigor and cost of experimentation. Nonexperimental designs have the added advantage of applicability after the fact.

Time Series Designs. Time series designs enjoy widespread popularity with evaluation specialists because of ease of application and the variety of data for some programs and because the method may be applied after a program has begun. The design can be illustrated as follows:

O O O X O O O O

The Os indicate that a number of observations on program progress are made before the experimental program is instituted or the ongoing program is adopted. In reality, the preobservations are gathered retrospectively, utilizing program records and other available statistics.

The reason for several observations before and after the introduction of X is the need to rule out episodic up-and-down swings in statistics as they occur on a random basis. If the researcher is confident about the stability of the indicators, he or she may wish to conduct a simple before-and-after test, which is illustrated as follows:

O X O

However, the researcher can have much more confidence in the data by taking the additional measurements.

The time series methodology can be illustrated using the example of an innovative patrol program in a hypothetical police department. The evaluator could obtain data on various measures for several years or months before and after introduction of the new system. Time series measurement of crime rates, citizen complaints, speed of response to emergency calls, and conviction rates due to citizen cooperation could be gathered. The latter three measures would be subject to the existence of record-keeping systems for such data.

A key problem lies in interpretation of the data and knowing when to declare the program a success, a failure, or merely a program. Suppose that the evaluator gathered data for the six months preceding and following the introduction of the new patrol program. A plot of the data might resemble the following:

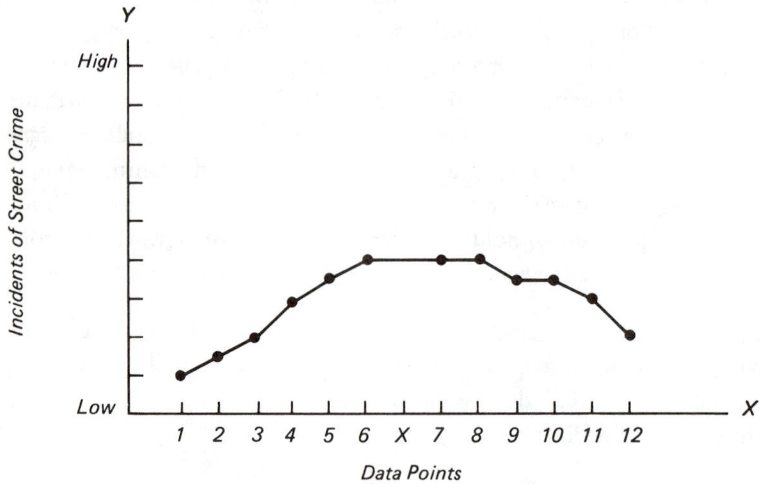

The incidents of street crime are represented on the *Y* axis of the above illustration. The *X* axis represents the various observations (data points) over the twelve months. The introduction of the experimental program is represented by an *X* between the sixth and seventh data points on the *X* axis.

The illustration shows that after an initial period of leveling off, street crime in the district dropped. The researcher cannot be sure, however, if the drop is a result of the program or history. History can be controlled by the introduction of additional time series data from other precincts, a neighboring city, or cities of comparable size and demography, or by comparing the pilot precinct with national statistics. The researcher's best bet is to gather data on other precincts or to compare the findings with the national statistics. Other cities of comparable size, for example, may not be keeping monthly records or may have introduced programs that could have an impact on the findings.

The multiple time series design can be illustrated as follows:

O O O X O O O
O O O   O O O

Assuming that the researcher decides to compare the experimental precincts with national figures, the plot might resemble the following:

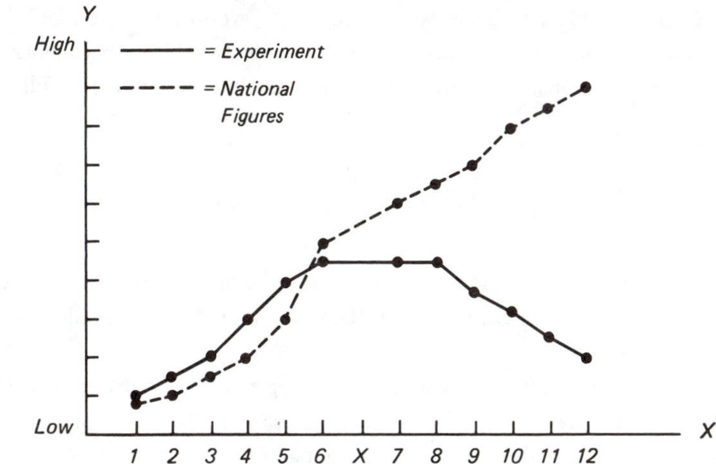

The multiple time series findings indicate that the street crime rate in the experimental district changed from being slightly higher than the national average for this category of crime to slightly below the precinct's own pre-program average. The drop in relation to the national figures indicates considerable change, because the slight decline in the local rate occurred during a period when national crime figures continued to rise.

Once the evaluation is complete, it must be determined whether the program was successful. This is done by the manager who assesses the indicators (such as crime statistics, citizen complaints, and so on) in the aggregate. Unfortunately, citizen complaints, emergency response rates, and the rates of street crimes cannot be forced into a one-dimensional measure of the program. The problem of interpretation might be further compounded if some indicators showed no change or if the program had an effect that was the opposite of that anticipated. For example, the evaluator might learn that the rates of such street crimes as muggings, assaults, and purse snatchings were reduced by the program, but that the number of family disputes, burglaries, and homicides had not been affected. And the evaluator might learn that the response time in emergencies had been cut by the program but that the overall conviction rate went down.

The above data are mixed. Maintaining evaluator neutrality is particularly important when the findings are mixed, because program advocates and opponents will both attempt to enlist the support of the evaluator. The function of the evaluator is to present the data in a straightforward, understandable format. Decisions regarding program continuation, expansion, or cancellation are the province of management.

Nonequivalent Control Group Designs. Nonequivalent control group designs are another variation of nonexperimental evaluation strategies. They compare the results of the program to similar organizations, groups, and the like. No attempt is made to make random assignments of subjects to the experimental and control group conditions. Nonequivalent designs can be diagrammed as follows:

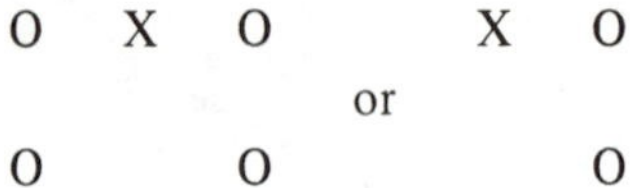

The diagram represents the pretest, posttest, and the posttest-only designs. Conspicuously missing from the diagram are Rs, which would signify random assignments to the conditions.

The use of nonequivalent control group strategies can be illustrated with the example of the alternative patrol pattern. Suppose that the police chief wished to know the impacts of the alternative patrol pattern on citizens' perceptions of the police and officers' perceptions of citizen cooperation. In this case, the posttest-only design would be preferable to the pretest and posttest design for three reasons: First, the additional data gained from the pretest might be only marginally beneficial when the additional costs are considered; second, the pretest step would necessitate ruling out the impact of testing on the subjects; and third, a pretest-posttest design would require that the evaluators posttest the same subjects who were surveyed in the pretest, many of whom may have moved in the interim.

Carrying out the posttest-only design would involve construction of two survey questionnaires to measure citizen and police perceptions. The citizen questionnaire would be used to survey a randomly selected sample of citizens living in precincts where the experimental program was implemented. The evaluator might also wish to gather a separate subsample of merchants in the experimental area, because they are disproportionately the victims of certain categories of crime and consequently experience a disproportionate amount of contact with the police.

The nonequivalent control group for the citizen/merchant survey might be a random sampling of citizens living or operating businesses in a roughly equivalent precinct, or the evaluators could sample citizen/merchant attitudes in the city generally. Police perceptions could be measured by sampling officers in the experimental precinct and a control precinct. Another, equally valid approach would be to survey all the officers in the experimental and control precincts.

Analyzing the data would involve comparing the mean responses of the citizen/merchant sample living in the experimental precinct with those of the control group. Similar comparisons would be made between officers in the experimental precinct and their control group. The evaluators would interpret substantially better attitudes among the experimental officers and citizen/merchant samples as a positive indicator of program success. Conversely, better attitudes among one or both control groups would reflect negatively on the program.

Matched-Pair Designs. Matched-pair designs fall into the category of nonexperimental designs because they do not rule out the effects of sampling, history, and maturity to the degree that experimental designs using random assignments do. The matched-pair approach assigns subjects to various experimental and control conditions on the basis of some common characteristic that the evaluator wishes to measure. In education evaluations, the characteristic frequently is the grade in school and/or the performance of the subjects on some other indicator.

Suppose that the experiment called for testing a new reading program. The researcher might proceed by assigning the children to the experimental and control groups on the basis of year in school and reading level. By so doing, two factors that could influence the findings are controlled. Other factors, however, could affect the outcome, such as home environment, classroom atmosphere, individual intelligence, and learning disabilities. Matched pairs may be preferable to nonequivalent control groups, such as the teacher to whom the student is assigned or the school attended. Complete random assignment, however, is the only way to assure maximum internal validity.

In some circumstances the matched-pair design is applicable to noneducational programs. Suppose one wished to assess the efficacy of a supervisory training program. Suppose further that the goals of the program were to improve organizational efficiency and employee morale through improvement of supervisory techniques. The evaluator might choose employee attitudes toward their supervisors as the measure of morale. The units of work produced per unit of cost might be the measure of productivity. The evaluator might also wish to determine whether the impacts of the training program varied with the level of responsibility assumed by the participants. Addition of the responsibility question makes it possible to compare the matched-pair supervisors who attended the training program with their peers in the hierarchy who did not attend.

The first step would be to select from among program participants and their peers a sample that was stratified according to position in the hierarchy. For example, if the training program was in a federal agency, the Government Service rating of the participants would be a rough indicator of their level of responsibility. Thus, if the sample of program participants contained two managers at the rank of G.S. 11, the researchers would then select two G.S. 11s who had not undergone the training for inclusion in the control group. The process would be repeated for each G.S. level until matched pairs of managers from each level of the hierarchy had been achieved.

The next steps would be administration of the employee morale questionnaire and assessment of unit productivity in relation to costs. The data would allow comparisons between program participants and their peers in the organization. Management could also compare the impacts of the training between various levels of the organization.

The findings of the above evaluation might be questioned–especially if program participants were far superior to the control group–because the evaluation did not rule out the possibility of "creaming" in the initial selection process. Creaming in this case would mean that participants had been selected on the basis of previous outstanding performance. If creaming had occurred, the evaluation could not determine which portion of the superior performance was attributable to the training program and which was attributable to the native abilities of the persons selected for the program.

Creaming is a threat to the validity of the training program, and therefore also a threat to the validity of the evaluation regardless of the design strategy employed. It could be ruled out merely by interviewing the training staff and upper-level managers before the evaluation began, in order to ascertain the selection criteria used in the training program.

THE PARTICIPANT-OBSERVER APPROACH

Examples of the participant-observer research design in process evaluation are presented in Chapter 6. The discussion here is confined to the method and its internal and external validity.

The participant-observer method is used to assess process. But process may be a goal of the program. For example, an education program might be established to provide teachers with training on the latest teaching techniques. The goal is therefore to produce better teaching methods in the classroom and, implicitly, better performance by students. Before student performance can be improved, however, the methods taught in the training program must be implemented in the classroom, and the only way to assess the application of the methods is to go to the classroom and observe.

The measurement instruments of participant-observer evaluations are the professional insights of the observer. The observer might also have a set of objective criteria on which the subjects are to be rated. The observer would visit the classrooms of teachers who had undergone training and assess them on the basis of the objective criteria. Additional reliability could be gained by having two or more observers rate each teacher independently using common criteria. These independent observations could then be combined into a single assessment of the teacher's overall performance.

The validity of the findings of participant-observer methodology cannot withstand rigorous questioning regarding instrumentation or the biases of the raters.

Whether the findings can be generalized beyond the organization studied is also questionable in a statistical sense. The strength of this methodology lies in its validity for addressing problems that program managers want solved. In the above example, the participant-observer method would determine whether the new techniques were actually being practiced.

Acceptance of the findings from participant-observer methodology rests in part on how the program staff perceive the evaluators' expertise. For example, if a nationally known educator attests that in his or her professional judgment an education program is an unquestionable success, the program staff is very likely to accept the judgment. More important, such expert testimony before school boards or legislators would probably carry the same weight as empirically gathered data. An expert assessment of program failure would also carry a great deal of weight. Generally, however, if the expert is retained by the agency he or she will advise as well as judge, and thus the program staff can make adjustments based on the expert's advice in order to bring program procedures into line with program goals.

MIXED DESIGNS

Programs are rarely so unidimensional that they warrant a single research methodology. Nothing, moreover, precludes combining outcome and process methodologies in order to advise program staffs as well as to measure program success.

The basic patrol program is an example of how the two methodologies can be combined to good use. In conjunction with the program staff, the evaluators could define the goals of the program in a way that provides for a rigorous outcome evaluation. The evaluators could also provide feedback on an ongoing basis by accompanying program participants on routine patrol and advising program officials on whether the officers were implementing the program as written. Program officials could then adapt supervisory procedures to ensure compliance, or revise the program plan, or reconsider the entire program concept.

These midstream adjustments could be taken into account in the time series design by extending the number of measurements after the introduction of X. In the case of the pretest-posttest design, interim data could be gathered. The methodology is not pure, but if the program is a success, the adaptations are justified. Such adaptations would not be permissible in one of the experimental designs, regardless of whether the program was functioning as expected.

The range of available evaluation designs extends from experimental designs to the participant-observer approach. Combining methodologies, moreover, generally adds to the program relevance of the design. The order in which the various designs were presented here was in no way intended as an explicit or implicit judgment of their relative worths. Experimentation was presented first because the design discussion followed the section on evaluation validity and experimentation is the design least sensitive to the problems of validity. However, experimentation is not always possible or desirable—especially when the evaluation is undertaken after the program has been in operation for some time.

The variety of program evaluation strategies available and the complexity of programs now requiring evaluation makes it mandatory for the evaluator to acquire a full range of evaluation skills. The evaluator who insists on using a specific design to the exclusion of all others is like the repair person who chooses to work on a single brand of washer—he or she runs the risk of being very lonely.

FOR FURTHER READING

Bennett, Carl A. and Lunsdaine, Arthur A., eds. *Evaluation and experiment: Some critical issues in assessing social programs.* New York: Academic Press, 1975.

Bernstein, I., ed. *Validity issues in evaluative research.* Beverly Hills, Calif.: Sage Publications, 1976.

Blalock, H. M. *Social statistics.* 2nd ed. New York: McGraw-Hill, 1972.

Campbell, Donald T., and Stanley, Julian C. *Experimental and quasi-experimental designs for research.* Chicago: Rand McNally, 1966.

Caponaso, J. A., and Ross, L. L., Jr., eds. *Quasi-experimental approaches: Testing theory and evaluation policy.* Evanston, Ill.: Northwestern University Press, 1973.

Cook, Thomas D., and Campbell, Donald T. *Quasi-experimentation: Design and analysis issues for field settings.* Chicago: Rand McNally, 1979.

Poister, Theodore H. *Public program analysis: Applied research methods.* Baltimore: University Park Press, 1978.

Welch, Susan, and Comer, John. *Quantitative methods for public administration: Techniques and applications.* Homewood, Ill.: Dorsey Press, 1983.

Chapter 5

CONDUCTING AN OUTCOME EVALUATION

Some evaluation specialists believe that little can be gained by evaluating programs that are not experiencing problems.[1] The argument is that if everybody concerned with a program (including legislators, program staff, and program clientele) is satisfied, an evaluation is not warranted. The same experts also believe that little can be gained from evaluations of programs that lack clear goals.

Clear goals are absent from many programs because many government agencies or subunits of agencies are service operations. These operations may involve services to other agencies or subunits. In addition, the mission of a service agency or subunit may vary from day to day, and this makes output measures of goal attainment difficult. This volume began with the assumption that most government programs can benefit from an objective assessment of their operations despite the absence of self-evident quantifiable goals. Moreover, there are instances when evaluations must be conducted whether or not the program is controversial or has clearly quantifiable goals. For example, state and local officials who administer federally funded programs may find that evaluations of their programs are a requirement of grant continuation.

The discussion of how to conduct an outcome evaluation is divided into two parts. First the various issues that surround the evaluation process, such as selection of an evaluator, hidden agendas, and negotiation of an appropriate design,

[1] Chief among the proponents of this position is Carol H. Weiss in her classic text *Evaluation research: Methods of assessing program effectiveness* (Englewood Cliffs, N.J.: Prentice-Hall, 1972).

are covered. Then a model for how to conduct an outcome evaluation is discussed and applications of the model are presented. Examples are taken from the areas of community health and law enforcement. The discussion concludes with a summary model that is applicable to any program undertaking an outcome evaluation.

Outcome strategies are concerned with measuring program outputs. Another kind of evaluation, known as the process approach, focuses on how to deliver a program rather than on outputs. The two methods differ in terms of what is observed and how the observations take place. In order to avoid confusion, process evaluations are presented in a separate chapter (see Chapter 6).[2]

PREPLANNING THE EVALUATION

Program administrators are frequently aware of the need to evaluate at the outset of their programs. Retaining the evaluator and negotiating the evaluation design early can benefit both the administrator who wishes to demonstrate program effectiveness and the evaluator who wishes to use the most appropriate measures. Failure to preplan the evaluation constricts the range of available evaluation strategies and the number of available indicators of program success.

Persons engaged in evaluation research also frequently possess managerial and programmatic skills. Bringing the evaluator in at the outset allows the administrator to utilize all the evaluator's skills. For example, an evaluator with experience in managing similar programs could help the administrator avoid such problems as inefficient organizational structures or inadequately trained staffs.

The authors do not advocate use of external evaluation consultants exclusively. To the contrary, both internal and external evaluations can be beneficial, depending on the needs of the agency. The decision whether to use an internal or external evaluation team is a critical first step in the evaluation process.

CHOOSING BETWEEN EXTERNAL AND INTERNAL EVALUATION

Before the techniques of program evaluation were widely known, choosing an evaluator involved selecting a proposal from among those submitted by various evaluation specialists located outside the agency. More recently, evaluation

[2] The importance of the process approach is exemplified by the existence of texts that focus almost entirely on this approach. See, for example, Michael Quinn Patton, *Utilization focused evaluation* (Beverly Hills, Calif.: Sage Publications, 1978).

training has been incorporated into the core curriculums of most public administration programs. In addition, the Office of Personnel Management has developed training programs to equip career managers with basic evaluation skills.[3] The net result is an ever-growing pool of career public servants who can conduct in-house evaluations, so agencies do not necessarily have to contract for evaluation services.

When the consumers of the findings are external to the agency, however, an outside evaluation may enhance the credibility of the results. For example, Congress may require evaluation as a condition of program authorization, and state and local authorities may have to provide evaluation information to federal funding agencies routinely as part of the accountability system. In such circumstances, an external evaluation can lend an air of objectivity to the findings. The objectivity of outside evaluators is rooted in the fact that they have no vested interest in continuation of the program. External evaluations also may be particularly beneficial when more than one program delivery unit is to be examined. The objectivity of the outside evaluator can allay the fears of bias on the part of competing program units, which might perceive an internal evaluator as a central office "hit man" bent on making the unit look bad.

External evaluations may present a perspective on agency problems that had not previously been considered. And in agencies that do not have regular evaluation units, external consultants may be the most cost-effective way to acquire evaluation expertise short of hiring new staff or expending resources on training.

Internal evaluations may be preferable in agencies that maintain evaluation units or planning units staffed by persons with research skills. The first benefit of an internal evaluation is the evaluator's understanding of the organization. In this case the evaluator is privy to past program decisions that are the basis for current operations. Internal evaluators also are aware of the actual program responsibilities of various organization actors, and this can save valuable start-up time in getting the evaluation on-line.

An internal evaluator's knowledge of the organization may lead him or her to a correct identification of the reasons for program failures, but internal evaluators are vulnerable to retaliation from officials who are criticized by the evaluation. Few internal evaluators are brave enough to unequivocally identify managerial error as the cause of program failure. When managerial error is suspected as the cause of program failure, the wise administrator may wish to retain an outside evaluator.

---

[3] This text was originally developed as a training manual for a pilot program to train G.S. 13-15 federal managers in the techniques of program planning and evaluation. The program was sponsored by the Dallas Regional Office, Office of Personnel Management.

THE EVALUATOR AND MANAGEMENT: DEALING WITH HIDDEN AGENDAS

Selection of an evaluator is not a unilateral administrative decision. The evaluators, for their part, must be aware of the motivations behind the commissioning of the evaluation. External evaluators and, to a lesser degree, internal evaluators must both resist efforts to make them pawns of management.

For example, a number of hidden agendas may underlie the decision to undertake an evaluation.[4] The hidden motive most frequently encountered by external and internal evaluators is the program administrator's desire to demonstrate program efficiency regardless of true program performance. In such cases, an administrator might commission an evaluation and demand that the outcome of the evaluation be favorable to the program. Then the evaluator must decide whether to sacrifice his or her professional integrity for an evaluation contract. Program evaluators should also avoid involvement in evaluations that are commissioned for the purpose of injuring one or another program unit. In such cases, administrators are interested only in negative findings and in affixing blame for program failure on one or another predetermined office.

Finally, evaluators should be aware of the proclivity of some officials not to decide. A program's staff, clientele group, or legislators may demand changes when the administrator has no intention of considering the evaluation finding or of changing program practices. An example of such an agenda occurred in the housing authority of a large eastern city. Tenants of a housing project submitted multiple complaints about mismanagement of the project, personal abuse at the hands of housing authority officials, and lack of response by housing authority police to emergency calls received after dark. When they had no reply, the tenants went to the media.

The housing authority quickly commissioned a study of the alleged problems as well as of the practices and procedures of the authority generally. It retained a well-known local professor to conduct an unbiased analysis. (Experienced evaluators were available, but the professor had no previous evaluation experience.) According to the professor, the study was underfunded and he was spoonfed information favorable to the authority. The professor did not voice these complaints at the time of the initial report, which neither confirmed nor denied the charges of the tenants. The program managers had apparently accomplished their purpose, which was to buy time while appearing to act. The professor continued his investigation over several weeks at no cost to the authority, and his findings confirmed all the tenants' charges. A second report was submitted

[4] The agenda discussion presented here relies heavily on the discussion presented in Weiss, *Evaluation research*, pp. 11–18.

to the governing board of the authority and to the local media. The program managers maintained that the second study had no validity because it was neither authorized nor compensated by the authority.

The evaluator must guard against such hidden agendas. The external evaluator can sometimes uncover various agendas during the preevaluation negotiating process, provided the evaluator possesses the requisite interviewing skills. At that point, the external consultant can (1) convince the program manager to abandon the hidden agenda or (2) decline to participate in the evaluation. Internal evaluation staffs are in all probability aware of the various agendas that underlie the decision to evaluate. Because internal evaluators generally do not have the luxury of being able to decline to participate, they must move cautiously.

The temptation to seek an evaluation whitewash of the program may appeal to program managers for whom evaluation is a condition of external funding and who maintain internal evaluation units. The tendency in such cases may be to engage in self-evaluation in order to control the findings. Most administrators, however, are honest brokers who are convinced that the program can withstand external scrutiny and that an external evaluation would be more objective. In such cases, the role of the internal evaluation staff is to present the external option to management and then, if necessary, engage in an objective internal evaluation.

When the evaluator suspects that the evaluation is being commissioned to discredit one or another program unit, the external evaluator may decline to participate. As an alternative, the external and internal evaluator can take pains to provide objective information on program success and the performance of various program units. If the unit in question is performing adequately, the evaluation may persuade the administrator not to move against it; if the unit is not performing, an objective evaluation would provide documentation to support administrative actions.

When an evaluation is commissioned to avoid or postpone needed changes in a program, the best course for the external evaluator is to decline participation. The internal evaluator, on the other hand, should provide as objective an evaluation as possible without compromising his or her professional integrity. If the administrator insists on changes in the evaluation strategy to predetermine the outcomes, or if he or she demands that evaluation findings be altered after the research is completed, the internal evaluator should consider whether to resign from the organization. Like their external counterparts, in-house evaluators must adhere to the ethics of their profession.[5]

---

[5] The role of the evaluator and the degrees to which he or she should allow administrators or others to influence the design has been a subject in the literature of evaluation for some time. See, for example, the collection of essays in Kenneth M. Dolbeare, ed., *Public policy evaluation,* vol. 2 (Beverly Hills, Calif.: Sage Publications, 1975), especially the essays by James S. Coleman and Ronald W. Johnson. See also Carol H. Weiss, ed., *Evaluating action programs: Readings in social action and education* (Boston: Allyn & Bacon, 1972).

The decision to commission an evaluation and the selection of an evaluator are preevaluation activities. The administrator and the evaluator, moreover, must reach agreement on how the evaluation is to proceed. These negotiations cannot be taken lightly, lest the evaluator become an instrument of management's bad intentions. Assuming that an acceptable understanding is reached, the next step is the design and conduct of an objective outcome evaluation.

DEFINING PROGRAM GOALS

There is no single perfect evaluation strategy with which to assess all programs. The strategies most frequently treated in the evaluation literature deal with program outcomes. Outcome designs seek to measure the degree of consistency between program outputs and program intent. Ideally, the success of public programs would be measured by the degree to which an agency's outputs have an impact on their environment which is consistent with the *intent* of the legislation that authorized the program. The theoretical or legislative intents of programs are generally expressed in such terms as improving the quality of life, making the community safer, or improving the environment, but evaluators find that program officials frequently cannot agree on a concrete measurable definition of program goals. Translating broad legislative intents into measurable program goals is the difficult first task in the design of an outcome evaluation.

A lack of goal consensus among program officials may be the result of different perspectives on the program. Persons at the top of the hierarchy are interested in broad questions of resource allocation and program impacts, but line personnel are concerned with the direct impacts of day-to-day activities and may not have given much thought to overall program goals. Because goal consensus is central to the execution of an outcome design, the evaluator must have consensus-building skills.

THE PERSONAL INTERVIEW

The evaluator can employ a number of goal-defining techniques. The most straightforward approach is the personal interview. Interviews may be conducted individually or in small groups, depending on the preference of the program administrator and the needs of the evaluator.

The initial contact between an outside evaluator and program staff may be with one or more members of the organization's planning or evaluation staff. Before proceeding with an evaluation, however, the evaluator should meet with the person commissioning the evaluation, usually the program's chief. The meeting with the program administrator should provide the evaluator with clear-cut guidelines as to the type of evaluation the person has in mind as well as the administrator's definition of program goals.

The evaluator should also meet individually or collectively with the program staff, or jointly with the evaluator and the staff. Follow-up meetings with program staff may be necessary if the evaluator senses that there is a lack of consensus regarding program goals which is not being dealt with in the meeting, or that the program staff is taking the company line and sublimating their own judgments to those of the administrator.

CONSENSUS-BUILDING ACTIVITIES

When a program staff and administrator are unable to agree on the goals of the program, the evaluator must engage in consensus-building activities. Goal consensus can be achieved by means of non-assembled decision devices, such as the Delphi technique, or by the group problem-solving techniques, popularized by organization development.

The Delphi Technique. Assuming considerable time availability and at least modest resources, the evaluator can conduct a Delphi survey.[6] The process begins by asking each participant to list program goals without consulting the others. The data are returned by mail to the evaluator, who combines the opinions of the participants into a single list of program goals. The list is fed back to the participants, who are then asked to assign a value to the various goals using an interval scale. The scale may be from 1 to 100, using all possible values or only gradations of ten, or the participants may be asked to evaluate the goals using a more simple scale of 1 to 10. The data obtained from this second round are collected and analyzed by the researcher.

Next, each participant receives information regarding the mean value assigned by the group to a particular goal, the range of values assigned to the goal by the group, and the value he or she assigned to the various goals in the previous round. The researcher may want to focus the attention of the participants by substituting the standard deviation of the group for the range of scores (the standard deviation reflects a range within which a substantial majority of the participants fall). Participants then are requested to reassess the weights they assigned in round two on the basis of the group data. If a participant's round

[6] The Delphi methodology was originally developed as a method of forecasting the future using panels of experts. For a readable summary of the Delphi methodology, see Barry Bozeman, *Public management and policy analysis* (New York: St. Martin's, 1979). The authors have also used the Delphi technique successfully in developing a consensus among middle managers with regard to a performance appraisal system to comply with the provisions of the 1978 Civil Service Reform Act. See Ronald D. Sylvia, "A Status Report on WPPR at the NASA Langley Research Center," unpublished report, 1977.

two value for a particular goal was extreme and he or she does not wish to modify previous judgments in light of the group data, that participant is asked to write a justification for the refusal, which is fed to other participants in the fourth round. The process can be repeated as many times as necessary in order to reach a general consensus on program goals. By the fourth round, however, three to five priority goals usually emerge from the group. The process can be shortened somewhat by asking subjects to both list and assign values to goals in the first round.

Group Problem-Solving Techniques. Time constraints and preferences of program managers may rule out use of the Delphi approach to consensus building. Instead, the evaluator may wish to employ one of the organization development group problem-solving techniques.[7] Group problem solving is so named because the technique involves bringing the relevant actors together in order to define the goals of the organization in small groups.

The first step is for the evaluator to explain the purpose of the meeting: to develop a set of agreed-on organizational goals for the purposes of program evaluation. The participants are then divided into small groups from six to eight persons and asked to generate a list of organizational goals. They are cautioned at the outset that the groups are to discuss various goals until a consensus list is achieved without voting. Voting may be democratic, but it allows group members to decide quickly based on limited discussion. The result of voting is usually a majority and a minority, and the latter may acquiesce to the group in the goal-negotiating process and later condemn the evaluation findings.

The next step is for the groups to report their list of goals. The evaluator should act as synthesizer of the group outputs while the groups are reporting. He or she should point out the goals that are consistent among the groups. A surprising level of consensus can be achieved by having small groups act separately.

If substantial consensus is achieved in the first round, the evaluator may choose to discontinue the discussions and proceed with the evaluation, or ask the participants to form small groups again and rank order the importance of program goals. If substantial goal consensus is not achieved in the first round, the evaluator may regroup the participants so that representative groups with diverse lists may work out their differences. To expedite the discussion, the evaluator may ask participants to discuss only their differences. The process may be repeated as many times as is necessary to achieve a consensus regarding program goals.

---

[7] For a complete summary of the group problem-solving techniques of organization development, see Wendell L. French and Cecil H. Bell, Jr., *Organization development: Behavioral science interventions for organization improvement.* 2nd ed. (Englewood Cliffs, N.J.: Prentice-Hall, 1978), especially pp. 117–138.

Like the Delphi technique, group problem solving can provide a set of consensus goals for an organization. And the group approach has the added advantage of allowing the evaluator to establish working relationships with the program staff. By running an effective goal definition meeting, an evaluator can demonstrate professional competence and build linkages to the staff.

THE IMPORTANCE OF GOAL CONSENSUS

Developing a group consensus can be time consuming, but the effort is justified if it results in a rigorously defined set of measurable goals. The evaluator should never try to define program goals unilaterally, no matter how well the evaluator believes he or she understands the program. Experience has taught evaluation specialists that program officials who cannot or will not reach a consensus regarding the goals of the program are in all likelihood the same persons who will criticize evaluation findings for measuring the wrong thing.

A MODEL FOR OUTCOME EVALUATION

After attaining a consensus on program goals, the evaluator must (1) determine the interrelationship of various program goals with program elements, (2) develop a set of indicators to evaluate the success of the program elements, (3) generate a set of valid measures to make the indicators operational, and (4) design the evaluation so that one can determine what the program outcomes are and whether they have been positive. In the latter case, the evaluator must examine unintended outcomes as well as intended outcomes. A model that illustrates the sequential interrelationships of the steps in the process is presented in Figure 5-1.

LEGISLATIVE INTENT AND PROGRAM GOALS

The legislative intent of the program, which can also be termed the theoretical goals, is usually stated in broad terms that are not readily quantifiable. Before an outcome evaluation can begin, the evaluator and the program staff must translate theoretical goals into quantifiable program goals. Defining program goals in close collaboration with program staffs, as discussed earlier in this chapter, can ensure staff cooperation with the evaluation effort and enhance the acceptability of the evaluation finding.

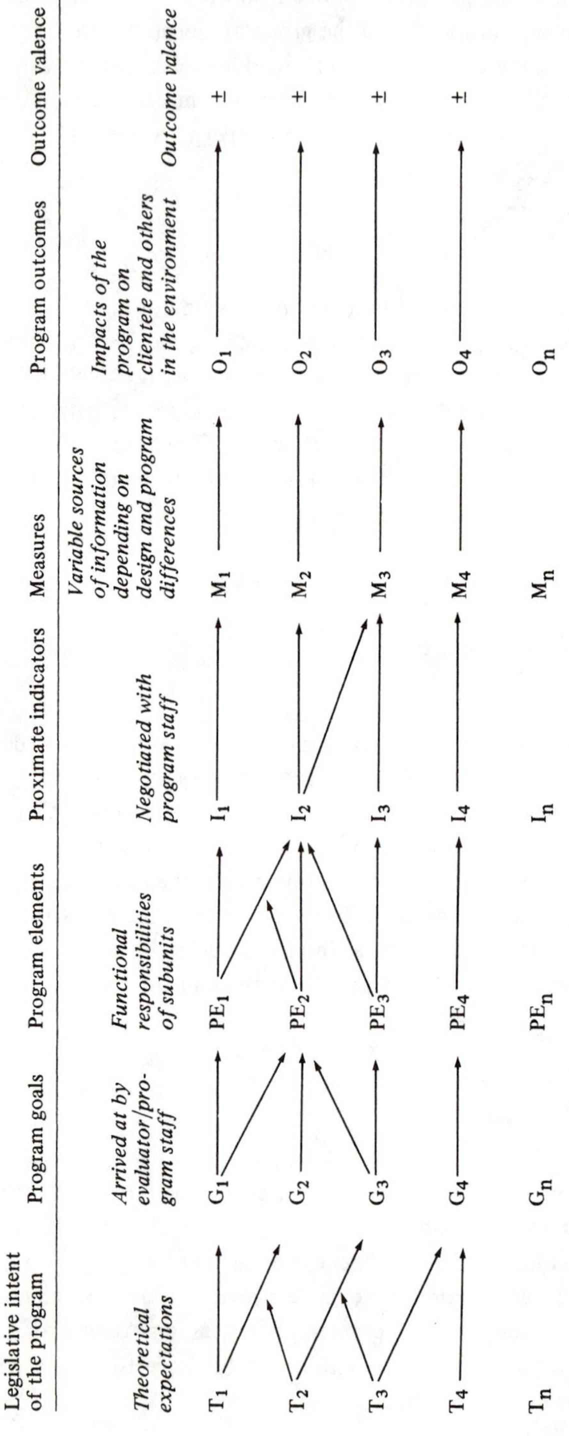

FIGURE 5-1
A guide to designing an outcome evaluation

PROGRAM ELEMENTS

The model also includes program elements, organizational subunits responsible for implementing the various components of the program. When evaluation findings are reported by program element, program managers learn which units need to modify their operations as well as how well the program is performing as a whole. Programs can be evaluated as a group without attention to the performance of specific units, but simple aggregate evaluations are more suited for consumption by legislators than by program officials.

PROXIMATE INDICATORS

Once goals have been defined and program responsibilities have been fixed on specific program elements, the evaluator can proceed with developing a set of proximate indicators against which program performance will be assessed. Like goal definition, identification of appropriate indicators is best accomplished in consultation with program staff. The range of available indicators is governed by the purpose and timing of the evaluation, the resources available for evaluation, and the preferences of the evaluator and the program staff.

MEASURES

Program measures reflect the operations that the evaluator will undertake to convert indicators to formal measures of program performance. They may involve comparisons of organization performance before and after program implementation. Measures may also be after-only measures of ongoing programs. Data gathered on the program may be analyzed singly or compared with similar organizations or program subunits that are not engaged in the program or that have implemented the program intent differently. Finally, measures may be based on existing records of the organization, on those of external accounting agencies, or on data generated specifically for the evaluation.

IDENTIFYING AND ASSESSING PROGRAM OUTCOMES

Correctly identifying and assessing program outcomes is the purpose of the entire strategy. Outcomes can be divided into primary and secondary categories: primary outcomes are direct program impacts, secondary outcomes are spillover effects that a program may have on clientele, other groups, or organizations. Program administrators are sometimes able to anticipate secondary outcomes. Whether or not secondary outcomes are anticipated, the evaluator must be aware of their potential and allow for them in the evaluation strategy.

OUTCOME VALENCE

Outcome valence is the final component of the model. An outcome valence may be positive or negative. A positive valence indicates that program efforts have achieved progress toward accomplishing program goals. The valence is negative when progress toward stated goals is achieved but agency clients or others are suffering negatively from the program in ways that were not anticipated by officials or by the evaluator. Programs that have no apparent impact can also be seen as producing a negative valence insofar as they are not performing.

APPLYING THE MODEL TO A COMMUNITY HEALTH PROGRAM

The preceding discussion of goal definition was, of necessity, generic. Here we use a hypothetical community health program to illustrate more concretely the steps in an outcome evaluation.

A community health program might have as its *theoretical goals* the upgrading and improving of community health and the quality of life. General health and the quality of life are difficult to quantify without translating theoretical intent into *program goals.* Therefore, the community health program might be divided into three program goals, dealing with locating and suppressing health hazards, providing diagnostic services to the community, and immunizing citizens against communicable diseases. The relationship between theoretical and program goals is illustrated in Figure 5-2.

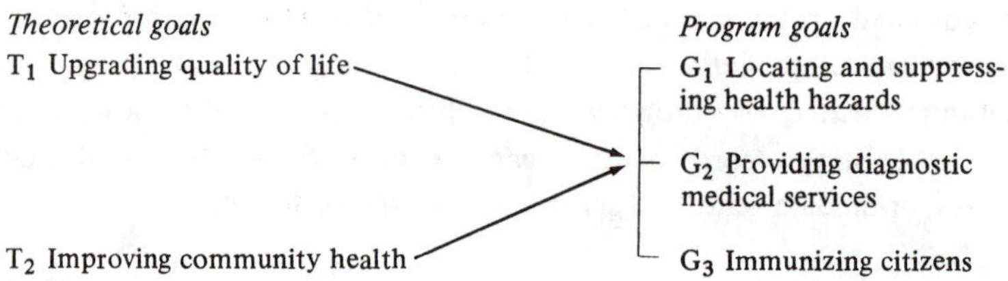

FIGURE 5-2
The relationship between theoretical and program goals

An arrow between a theoretical goal and a program goal indicates how the agency has interpreted the intent of the legislature. A theoretical goal that is linked to more than one program goal indicates that the former must be translated into more than one program goal.

Program elements are organizational subsystems organized to carry out one or more program goals. In the present example, the program elements might be a prevention and suppression unit and a system of diagnostic teams that operate health clinics and provide immunization services as necessary. Before illustrating more fully the use of program elements, we should point out that inclusion of program elements is not essential to the conduct of an outcome evaluation. For example, a community health agency's overall effectiveness could be evaluated without measuring the performance of specific program units. Program element analysis is most beneficial when the consumers of the evaluation are program staff members who are concerned about how well various units are performing. Inclusion of program elements in this discussion is an expression of the managerial-consumption biases of the design strategies included in this text.

In our health care example, the prevention and suppression staff might be made up of medical and environmental specialists who are responsible for finding and suppressing health hazards in the community. Such hazards might range from poor air and water quality to rats in concentrations that present a health problem. The diagnostic unit might be staffed with persons responsible for conducting health clinics where citizens could obtain medical advice and preliminary treatment of a range of health problems from communicable disease to prenatal care for expectant mothers. The diagnostic units might simultaneously provide immunization services for a variety of diseases. Special immunization projects could operate on an ad hoc basis when outbreaks of specific diseases occur, or at the start of school years, when large concentrations of children become available for treatment.

Because health care operations are complex and interrelated, the range of activities discussed above only scratch the surface of health care services. There are enough, however, to demonstrate the relationship between program goals and program elements (see Figure 5-3).

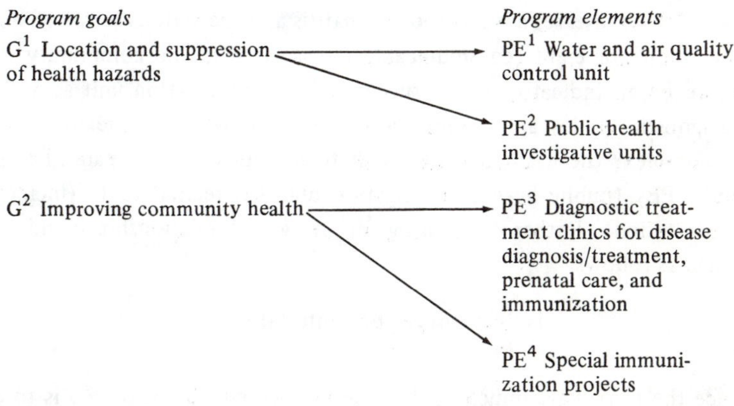

FIGURE 5-3
The relationship between program goals and program elements

In addition to temporary or experimental efforts, government programs are established to provide continuing services as well. Most programs achieve only partial success in alleviating their targeted problems. In the community health example, eradication of a disease through immunization of the public would indicate program success, but if the program reduced the rate of the disease by only 80 percent, one could not realistically label the program a failure.

The evaluation specialist has a number of indicators available for measuring program progress short of complete success. In our community health example, the *proximate indicators* of water and air quality unit success might be the incidents of water and air pollution that the unit identified during a specified period. Indicators of hazard prevention might include the instances of pest infestation reported by the hazard prevention unit and the volume of reported diseases attributable to pest infestation. The success of the diagnostic unit might be measured using the indicators of the amount of prenatal service provided and the amount of communicable disease diagnosis in which the organization engaged. The indicators of immunization services success might be the instances of communicable diseases in the community over time and the type or quality of immunization services delivered.

Secondary services to the public could also be a valid indicator of program success. For example, the evaluation might disclose that the prenatal clinics spent considerable time counseling expectant mothers with regard to available welfare and birth control services.

The arrows in Figure 5-4 indicate unidirectional linkages between program elements and proximate indicators. In reality the relationships are rarely so simple because program units engage in mutually supportive services. For example, the investigation unit might be responsible for locating persons who had come in contact with clinic patients treated for communicable diseases. Examples of these diseases are infectious hepatitis and the various forms of venereal disease. Indicator eight (communicable disease rate in the community) would therefore be an indicator of the success of the investigation unit as well as of the diagnostic clinics and immunization projects. When interrelationships become complex, the evaluator may wish to use subscripts in parentheses. For example, PE_2 (public health investigation units) is related to I_3 (incidents of pest infestation). Rather than using an arrow, the relationship could be diagrammed as follows:

I_3 Incidents of pest infestation
PE_2

Once the proximate indicators have been selected, the next step is to devise a set of *measures* to put them into operation. However, the measure selection process may result in a rethinking of the indicators because data is not available.

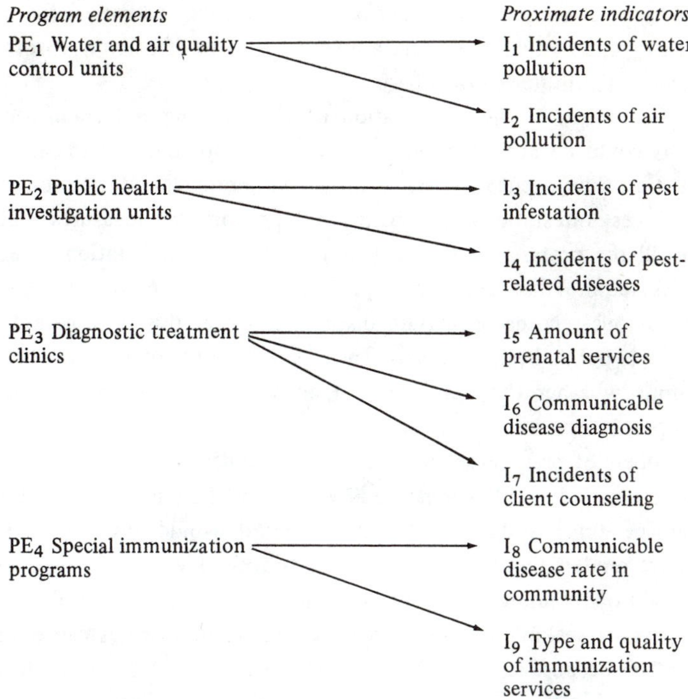

FIGURE 5-4
The relationship between program elements and the proximate indicators

Selection of an appropriate set of measures to put the indicators into operation is governed by the design employed and the timing of the evaluation. When the program under investigation is temporary or experimental, the evaluator can gather baseline data before the program is initiated. The measure of program success would be the changes resulting from the program as measured by a comparison of data gathered at the end of the program with the baseline data. Statistics on disease rates in the community gathered before and after implementation of an immunization program would be an example of this type of measure.

When evaluating an ongoing program, one may wish to utilize a combination of measures that are gathered from program records and ad hoc measures that are created solely for evaluation purposes. The latter frequently involve surveys of program clients and staff. Such surveys are popular because they are convenient and because a number of indicators can be put into operation in a single survey. Both records and survey measures can be illustrated using the community health example.

Air and water quality enforcement programs could be measured by collecting samples of air and water in various parts of the jurisdiction over time. When combined with a geographical analysis of enforcement activities, the air and water quality data would constitute a measure of how effectively enforcement activities were distributed in the district.

The effectiveness of the investigation unit in locating and eradicating pest infestations could be accomplished by a review of program records and health care statistics. The agency's own records could provide information on the number of pest infestations investigated by the unit, the measures that were taken to kill the pests, and the frequency of reported infestations over time. Most states require that health care providers, such as doctors and hospitals, maintain copious records on certain diseases and that this information be reported to local public health officials. These records could be reviewed over time to determine whether there had been a change in the incidence of diseases related to pest infestation.

The numbers of prenatal service clients, communicable disease diagnoses, and the clients counseled could be measured by a review of clinic records and surveys of program personnel and clientele. Clinical records would reveal what types of services were rendered and whether expectant mothers were utilizing the health clinics in lieu of retaining private physicians. Clinic records might also indicate whether persons seeking treatment for communicable diseases came on their own initiative, because someone had advised them to, or because the investigative team located them and asked them to seek help at the clinic.

Members of the clinic staff could be surveyed to determine how they allocate their time, their perceptions of the quality of services rendered by the clinics, and what suggestions they have for upgrading services at the clinics. Clients could be surveyed as to the quality of service they received at the clinics, how they were treated by program staff, and what referrals they received from the staff.

Measuring the success of the immunization programs might involve a review of records for the jurisdiction and available state and national data. The evaluator might review the records for several years prior to and after the introduction of various immunization programs. The time series data on state and national disease trends would indicate whether changes in local disease patterns could be attributed to immunization programs or to a general change in the occurrence of the disease.

The relationships between indicators and the various measures (see Figure 5-5) can be straightforward, one-to-one relationships, but sometimes they involve complex relationships, as when the success of clinical services is assessed by three separate indicators that are put into operation by four measures. The measures, in turn, may serve multiple purposes. Measure six (the staff survey) would measure the quality of prenatal care, the incidents of communicable diseases diagnosis, and the amount of counseling service to clients. The survey would also gather staff perceptions regarding changes necessary in clinical services.

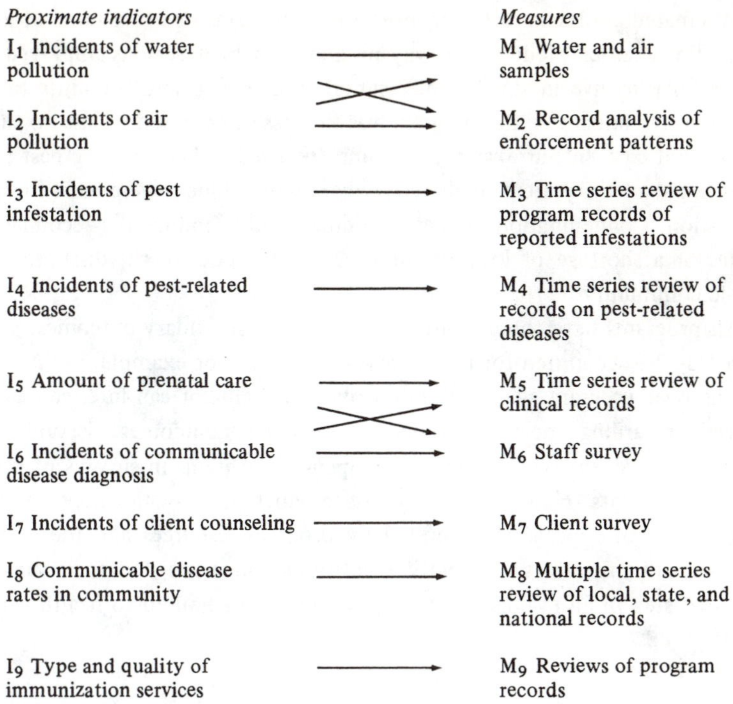

FIGURE 5-5
The relationship between proximate indicators and the measures

The next step in the evaluation strategy is to consider the interrelationships between measures and ***program outcomes.*** Because program outcomes can be either positive or negative, the evaluation should provide for either *outcome valence.* When the data sources are such things as health statistics, either valence may appear in the analysis. When the evaluator is employing survey questionnaires, he or she must construct the instrument so that it probes for negative as well as positive program results.

A program is thought of as producing a positive valence when satisfactory progress toward a stated goal is accomplished. Negative valence occurs when the program produces results that are in the direction opposite from that intended. Programs that produce no apparent change can also be thought of as negative.

Most direct outcomes can be inferred from the goal definition phase of the strategy, but not all outcomes can be anticipated. Programs may also produce secondary outcomes that can be either positive or negative. In our community health care example, a positive secondary outcome would be the counseling of expectant mothers with regard to the availability of social services. Secondary outcomes may be new problems that stem from a successful program. Such unanticipated problems often involve other programs or other units of government. For example, federal highway programs enabled persons employed in the central city to live in single-family suburban dwellings and commute to work. The negative impact of the program was the loss of the middle-class taxpayer to the central city. In our example, the unit involved in investigating pest infestations might identify areas of the city where whole blocks were unfit for human habitation. Condemnation of these buildings could produce the secondary outcomes of a shortage of low-cost housing for the poor or the disruption of an ethnic community.

All programs have the potential for producing secondary outcomes. Some of these can be accounted for in the design strategy. For example, during consultations over program goals and indicators, the evaluator can interview program officials regarding their knowledge of secondary outcomes. Secondary outcomes can also emerge as a result of open-ended items in surveys of program staffs and clients. However, the extent to which an evaluator may assess spillover effects of a program is limited by program resources and the desires of program administrators. Correctly identifying outcomes and their valences is the final step in an evaluation strategy which is not limited to health care programs.

APPLYING THE MODEL TO AN EXPERIMENTAL POLICE PROGRAM

In Chapter 4, experimental police programs were used to illustrate time series designs. The discussion now returns to that example in order to illustrate the versatility of the evaluation strategy and the variety of indicators and measures available to the imaginative evaluator.

The turbulence of the 1960s produced a plethora of federal programs aimed at upgrading and improving law enforcement at the local level. The theoretical goal of these programs was to produce safer streets. Government assistance to local police departments was earmarked for purchase of equipment, training of officers, and development of innovative police service strategies. A number of cities received grants for developing alternative patrol strategies to reduce crime through better identification of criminals and closer cooperation between police officers and the communities they served. The projects were variously known as "operation neighborhood," the basic patrol plan, and team policing.[8]

---

[8] For a complete report on an evaluation of one of these programs, see Peter B. Bloch and David I. Specht, *Evaluation of operation neighborhood* (Washington, D.C.: Urban Institute, 1973).

These innovative programs were alternatives to traditional patrol patterns in which officers were assigned to a district at the start of a shift without regard to whether they were familiar with the neighborhood's geography or population. Patrol cars were dispatched to emergencies on a first-call, first-serve basis. A patrol car assigned to one neighborhood might be dispatched to the other end of the precinct because that district's patrol car had been dispatched elsewhere. Under the traditional system, the immediate need of the person calling the police was met; the needs of citizens in the unpatrolled neighborhood for on-going protection and immediate response were held in abeyance.

The second purpose of the experimental neighborhood operations was to break down barriers between officers and the communities they served. Divisions between the police and citizens resulted from the ad hoc patrol and dispatch assignment procedures. Citizens complained about harassment by officers and officers' failure to distinguish between criminals, potential perpetrators of crimes, and law-abiding citizens destined to be the victims or witnesses of crimes. For their part, officers complained of citizen hostility, aggression, and non-cooperation with investigations.

The *theoretical goal* of upgrading law enforcement in order to achieve safe streets was translated into the *program goals* of improving police patrol effectiveness and upgrading police community relations. The *program elements* were an alternative patrol pattern and a systematic attempt to establish one-to-one relationships between patrol officers and the citizens they served.

The experimental patrol pattern involved assigning tactical groups of officers to a neighborhood on a repetitive basis. Command of these officers was usually vested in an area commander below the precinct level. Patrol cars assigned to an area were not to be dispatched to the outside except in dire emergencies.

The officers-citizen relations program involved giving officers additional training aimed in part at upgrading their communications skills. Other aspects of the training stressed the importance of good citizen-police relations in securing citizen cooperation with police investigations and citizen willingness to testify.

The *proximate indicators* of program success include changes in the crime rates in the experimental precincts, changes in the response time of patrol cars to emergency situations, and changes in citizen and police attitudes with regard to police community relations. A final indicator of program success would be changes in the conviction rates of criminals whose crimes occurred in the experimental precinct.

An appropriate *measure* of crime rate changes would be a multiple time series comparison of crime statistics collected in the experimental precinct and a control precinct. Data from the experimental district could also be compared with national or statewide figures. Response times on emergency calls could be measured by having dispatchers log the data on the elapsed time between the dispatching of patrol cars and their arrival at emergency scenes in the experimental and control precincts.

Changes in police-community relations could be measured using a survey of police-citizen attitudes in the experimental and a control district. A time series study of conviction rates as a result of citizen cooperation between the experimental and control districts could be used to measure whether the community was responding to the new programs.

The evaluator might also wish to measure the portion of change in conviction rates that could be attributed to changes in patrol patterns and to increased citizen cooperation. A review of criminal proceedings would indicate which arrests and convictions resulted from officers' catching perpetrators in the act and which resulted from citizen reports of crime or citizen cooperation with follow-up police investigations.

The alternative patrol pattern experiment also can be used to illustrate how *program outcomes* may vary from those anticipated by the evaluator. The patrol portion of the program was intended to make officers more readily available for quick responses to emergencies and more knowledgeable about the neighborhood and the citizens who lived there. The community relations portion of the program was intended to make it possible for officers to come in regular contact with citizens. An assumed outcome of this contact would be improved citizen attitudes toward the police, which in turn would result in closer cooperation. Fewer complaints of police harassment were also expected as a result of police efforts.

Officers in the experimental precinct might have accepted only a portion of the program goals. Suppose that they accepted the notion of team patrol, enhancing their knowledge of the neighborhood and the reduction in response time elements of the program, but did not accept a requirement that they engage in walking patrols and one-to-one contacts with citizens. In that case, the evaluation might report a reduced rate in the category of street crimes, such as muggings and purse snatching, but citizen attitudes toward the police would remain unchanged. The frequency of complaints regarding police harassment might actually increase.

A reduction of street crime and increased harassment complaints could both result from officers' increasing their use of aggressive tactics, such as stop and frisk (stop and frisk involves detaining suspicious persons and searching their persons for weapons and contraband). Stop and frisk might be a natural by-product of team patrolling of neighborhoods because officers would be better able to identify suspicious persons. Even when they are not arrested during a stop and frisk encounter, persons who participate in street-corner society are likely to go elsewhere in the face of aggressive police tactics. Both recipients of stop and frisk and law-abiding citizens who witness the encounter are likely to complain about police harassment.

Stop and frisk activity was not an element in the program—it was an unanticipated secondary outcome. In light of the outcome, police officials might wish to weigh the benefits of stop and frisk activities (fewer loiterers and reduced crime rates) against the negative spillovers (potential violations of civil liberties and negative citizen attitudes toward the police). The community relations aspects of the program might also be reconsidered, in light of the above, for modification or abandonment.

The outcome strategy presented in this chapter is broadly applicable whenever the purpose of the evaluation is to measure program impacts. In order for this or any strategy to succeed, however, the evaluator must be sure that each step of the strategy is carried out in a systematic manner. Finally, outcome evaluations are most effective when the consumers of the finding are external to the organization. Evaluation consumers are most frequently program officials who are interested primarily in how the mission is being accomplished. These officials might commission an evaluation to help them clarify program goals or to assess how program elements are carrying out their mission. These types of evaluations are commonly known as process evaluations and are the subject of Chapter 6.

Case Studies

A HYPOTHETICAL CASE: PROGRAM EXPERIMENTATION

In the late 1970s one state sought and received Department of Labor funds to experiment with an alternative job placement program for welfare recipients that cost more but that held a promise of more meaningful work, which in turn might lead to longer-term job retention. The new program called for work-readiness training for welfare recipients and increased benefits to the employer.

The training program required that the staff develop a program to make welfare recipients more work ready. The program was to provide training in resume construction, job-interviewing techniques, and conducting systematic job searches. It was also to make participants aware of the problems and frustrations facing a single parent who works. Finally, the staff was to approach employers and explain the program in order to increase the number and quality of potential placements.

The principal modification of the program vis-a-vis the employer was a change in the level of benefits and the formula for payment. Under the experimental program, employers would receive 75 percent of the costs of hiring and training welfare recipients, a 50 percent increase over current programs.

Payments in the experimental program, however, would vary as the employee developed. During the first few weeks of employment, the government would pick up 100 percent of the employee's salary, then the amount went to 50 percent, then to 25 percent.

The reasoning was that employers might be more inclined to hire and train program participants if aid were increased during the initial training period of employment when the employer's outlay was at a maximum. When the increased incentive was combined with better-equipped participants and extra efforts of the program staff, the net result should be long-term placement in meaningful work and ultimately a reduction in the welfare rolls.

Federal and state program officials did not agree about what the relative importance of the program's components were or should have been. Federal officials were inclined toward the employer incentive element, while state officials were convinced that making the participants more work-ready and helping them search for meaningful work was the primary benefit of the program.

In order to maximize the validity of the test results, the program was to be run on a pilot basis in two cities. City A was largely a manufacturing center engaged in heavy industry that employed low-skilled workers in assembly-line jobs. The unemployment rate was seasonably variable, with higher unemployment occurring in the winter months. City B was the state capital, where the state university also was located. Employment in this city was primarily white collar or service oriented. Unemployment rates for City B were stable and low. Workers in City B, however, were compensated at a lower rate than workers in City A.

The initial funding grant provided for an experimental program of five hundred participants to be divided any way the program saw fit. The principal purpose of the program was to determine whether or not the combination of the new incentives and job-readiness training worked better than traditional programs. The emphasis was therefore on designing the evaluation at the time of program start-up. An outside evaluation approach was chosen.

Assume that you were an employee of the Bureau of Government Research at Ivory Tower University. The bureau contracted to perform the evaluation. The bureau head knew about the different perceptions that federal and state officials had of program goals and was understandably concerned that the evaluation strategy meet the needs of both audiences.

Develop an evaluation design to test the effectiveness of the program in meeting its goals:

1. Begin by using the evaluation model presented in the chapter to define the theoretical intent, program goals, program elements, proximate indicators, and measures. Then illustrate your design using an appropriate array of Rs, Xs, and Os as presented in Chapter 4.
2. Is it worthwhile to attempt to achieve goal consensus among state and federal officials? If so, what strategies would you pursue to achieve consensus?

3. Design your experiment to ensure a proper evaluation and devise an appropriate evaluation design.
4. How will you go about selecting program participants to ensure that there is a minimum of bias built into the evaluation design?
5. What kinds of measures are available for testing the impacts of the culture of poverty on the participants and control groups? Is there a difference between welfare people and the general population with regard to their attitudes about the value of hard work?
6. What kinds of measures will you use to determine whether the additional incentives to employers are the principal factors in program success or failure?
7. What kinds of indicators will you use to see whether there is a difference between the experimental program and regular WIN programs in terms of placement in meaningful work?
8. What kinds of measures will you use to test the preemployment training component?
9. How and when will you know if the program was a success?

A HYPOTHETICAL CASE: EVALUATING A TRAINING PROCESS

Each year the National Aeronautics and Space Administration (NASA) cosponsors a training program for college professors with the American Society for Engineering Education (ASEE). One stated goal of the program is to acquaint an interdisciplinary team of researchers with skills in systems design. The hope is that these professors both from engineering and from the social sciences will return to the classroom and teach their students systems design techniques. By so doing, NASA and the ASEE hope to broaden the pool of persons in the society with these skills, thereby benefiting the engineering profession, NASA, and society as a whole.

A second goal of each annual project is to solve a problem that is of interest to NASA. Teams may work on such things as designing an offshore airport, the applications of aerospace technology to agriculture, or the design of transportation accesses to international airports. The idea is to tap the skills of the professors as well as equip them with new skills.

Assume that you are the project director and that for the last three years the process has gone as follows: Twenty professors are brought to one center for ten weeks. There they are briefed on the resources available, for instance, computer access and library resources. After the one- or two-day orientation, the team settles into its ten-week work assignment.

The team is divided into small groups of five members from a variety of disciplines. Each group is to work independently of the others and to define the elements in the systematic resolution of the current year's problem, a phase that takes about two weeks. Then the groups report their recommendations to the larger group, and a consensus on direction is arrived at with the entire design team working as a single unit. This portion of the deliberations may take from one to two weeks, depending on the diversity of the various groups' recommendations.

Once consensus is reached, the professors are divided into smaller work units, which take responsibility for various phases of the design project. The remainder of the project involves independent work on the part of the professors and small groups. An editorial board is responsible for synthesizing the productivity of the subgroups in a final report, which is presented to the NASA officials.

Assume that you, as project director, wish to determine, along with your staff:

1. How effectively the process has taught systems design to the professors
2. How effectively the design team experience has utilized the skills of the various participants
3. What phases of the process could be done differently, and how these differences might affect the professors' perceptions of project success

Assume further that you are drawing near to the end of a design team project and want to experiment with changing the design team experience to use the ten-week period more efficiently and to cut down on anger on the part of participants who do not believe that their small-group outputs were treated fairly by the larger group.

1. What changes will you make in the processes the next design team will undergo?
2. In order to develop baseline data on which to assess the merits of the changes you are making,
    a. What outcome measures will you use?
    b. What elements in the process will you assess?
    c. How will you assess the current year's process now that it is nearly complete?

A HYPOTHETICAL CASE: MIGRANT WORKERS

Each year there are a number of accidental pesticide poisonings of migrant farm workers. The actual number of cases is difficult to determine because the symptoms (headaches, nausea, vomiting, and so on) are common to other ailments, such as flu. As a result, people may or may not know that their illness is pesticide-related. The number of documented cases, however, is sufficiently large to warrant governmental countermeasures. Officials of the Environmental Protection Agency (EPA) and the Department of Agriculture (DOA) are convinced that the poisonings are mostly accidental and stem from improper handling of pesticide materials by workers. For example, workers may not be aware of the need to wash their hands after handling pesticides and before eating or using the restroom or to change work clothes each day when handling pesticides. Other poisonings are the result of workers' placing unused pesticides in milk cartons or other improperly marked containers.

The EPA and the DOA have agreed to co-sponsor a program to educate migrant workers about the hazards of pesticides. The EPA has agreed to provide $100,000 for a pilot program to administer the grant. The actual training will be provided by the Florida Department of Labor and be overseen by the Florida Department of Agriculture.

The grant specifies that the Florida Department of Labor will develop and pilot-test education modules aimed at preventing accidental pesticide poisoning. The program, which is to be developed in both English and Spanish, will initially involve two hundred migrant workers. Program development will be particularly difficult given the low levels of literacy among migrant workers.

The evaluation, to be conducted by the Florida Department of Agriculture, is to assess the impacts of the program and should include: (1) a method for assessing the quality of the education modules developed in the program; (2) a method for assessing the actual delivery of the training; (3) a method for measuring the success of the program in improving the knowledge of the participants regarding the correct use of pesticides and potential dangers of mishandling them.

Apply the evaluation model on page 112 to the experimental program:

1. Specify goals, elements, indicators, and measures that are appropriate.
2. Develop an appropriate design strategy using Xs and Os (see Chapter 4) to illustrate how you will proceed.
3. Discuss the limitations that influence the evaluations. What kinds of general program application inferences will you be able to make on the basis of this evaluation?

A HYPOTHETICAL CASE: TEACHERS' AIDES

The city of Sorgum, Arkansas, is a community of 25,000 located in northwestern Arkansas near the Oklahoma and Missouri borders. For the last fifty years the population of Sorgum has been engaged primarily in agriculture or in servicing the agricultural community. Five years ago, the major utility company finished construction on a nuclear power plant that services the needs of cities in three states. The plant now accounts for 54 percent of the jobs in Sorgum. Seventy percent of those employed at the plant are recent arrivals who have no roots in the local community. This population influx resulted in rapid construction of two new grammar schools (grades one through six) and one new middle school (grades eight and nine). The next anticipated pressure will be on the high school, which must expand its facilities to accommodate the maturing newcomers.

The district now has four grammar schools, two middle schools, and one high school. Some 24 percent of the students in the district are bused in from farms surrounding Sorgum. A pilot program for busing city residents also has been under way for two years as a result of threatened lawsuits by the local black and Native American communities, who feel that the current racial composition of the schools is inequitable. Generally, the minority communities are made up of long-term residents and not affected by the recent influx of population.

The racial composition of the schools causes community unrest. The high school is fully integrated. Problems at that level center on student assignments to the new and old facilities when the new facility becomes operational. The middle schools also are relatively balanced. Johnson Middle School is 75 percent white, 20 percent black, and 5 percent Native American; Kennedy Middle School is 34 percent black, 12 percent Native American, and 54 percent white. The actual racial composition of all school-age children in Sorgum is white 60 percent, black 31 percent, Native American 9 percent.

The principal problems of racial harmony stem from student distribution in the grammar schools of Sorgum. Jefferson School is located in a predominantly white part of the city. The student composition at Jefferson is 95 percent white, 3 percent black, and 2 percent Native American. The latter percentages are a result of the pilot busing program. Lincoln School is located in the black section of the city and is made up of 84 percent black, 11 percent Native American, and 5 percent white. The white students are bused to Lincoln School. The two newer schools have operated for three years and are located in areas accessible to all three communities. The composition of the student body at Carver School is 65 percent white and 35 percent black. Roosevelt School is 40 percent black, 40 percent white, and 20 percent Native American. Minority parents are as much concerned about the overcrowded conditions at the Lincoln and Roosevelt Schools as they are over racial balance.

The Sorgum school board is made up of seven members elected at large. The membership includes five men and two women. To date, no black or Native American has secured a seat on the board, although black and Native American candidates have made strong showings in the last three elections. The occupational status of the board members reflects the community they serve. The president of the board is a nine-term farmer. Two other rural residents also hold seats on the board; one is a retired teacher and the wife of a farmer in her eighth term, the other is a medical doctor who raises horses near town and who is in his third term. The remaining board members are a grocer, now in his second term; a housewife, now in her first term; an executive at the power plant, also in his first term; and a service station operator, who has been on the board for thirty years and is in his last term.

The membership of the board is deeply divided on the question of busing; the newer members see a need for the program, and the senior members strongly oppose such innovations. The swing vote on such issues rests with the doctor, who voted for the busing program in the interest of community peace. The board is also skeptical about ethnic studies programs and special programs for persons with learning disabilities, even though the presence of the power plant has considerably increased district revenues. The conservative members of the board would rather turn excess tax revenues back to taxpayers than sink them into high-cost, low-return programs for the educationally handicapped. They want the curriculum to consist of reading, writing, and arithmetic.

The vast majority of teachers in the district are Sorgum natives who attended one of the two state-supported universities for four years and returned home to teach. Most recently, however, an increasing number of the teaching positions are being filled by newcomers who are affiliated in one way or another with the power plant. Most notable among the new hirees are four black teachers who are not Sorgum natives, bringing the total number of black teachers in the district to twelve. The remaining eight black educators are longtime Sorgum residents, one of whom is principal of Lincoln School.

The school system is headed by a thirty-five-year-old superintendent now serving his second year who has a doctorate in education from a university in a neighboring state. He has spent the last eighteen months getting acquainted with the district and establishing himself with the board, the staff, and the community. He has acquired a reputation as an innovative, levelheaded man, but there is some resentment about his appointment which stems from his being selected over two local candidates, one the black principal of Lincoln School, the other the high school principal.

The superintendent has two assistants and a clerical staff of four to run the entire district. The void is filled somewhat by school principals who assume more duties than required by the normal course of school operations. This strategy has worked fairly well, but the superintendent is preparing to push for additional central office personnel in the upcoming budget year.

Two months ago, a notice from the U.S. Department of Education arrived at the superintendent's office specifying that an evaluation of the teachers' aide program which had been funded for a five-year period was overdue and that, if the district wished to apply for continued funding of the program, it would have to demonstrate that the goals of the program had been met. The last sixty days have been spent reviewing the original grant proposal and examining how the program was set up and operated.

A central office review of the program indicated that the former superintendent had indeed secured a grant for a teachers' aide program. Because Jefferson and Lincoln were the only two grammar schools at the time of the grant, all the aides had been placed in those schools, where they remained. The office staff vaguely remembered that the former superintendent, now deceased, had two motivations, first to preserve racial harmony in the district, second to run "some sort of experiment."

The Department of Education wants to know how well the program is achieving its stated goals. The program calls for an aide in the classroom (1) so that the teacher can spend less time on administrative duties and more on teaching, (2) so that classroom discipline can be improved, by the addition of another adult, and (3) so that students can receive more individualized attention in math and reading. All this was supposed to improve the educational experience and provide maximum educational benefits for the dollars spent.

The present superintendent has called a meeting of the two central office professionals and the principals of the four Sorgum grammar schools. The purpose of the meeting is to determine how to proceed in light of the following facts:

1. The district intends to reapply for the funding.
2. The superintendent wants to know whether there is a better way to distribute the aides.
3. There is $5,000 in a contingency fund, which could be expended on the evaluation, but the superintendent would prefer to spend the funds on books for the library.
4. It is hoped that the evaluation will demonstrate program success, which will support the request for additional funds from the board in the upcoming budget year.

Using the background provided in this case,

1. Define the audience of the evaluation.
2. Determine whether the evaluation should be internal or external.
3. Specify the goals of the program.
4. Devise an appropriate evaluation design.
5. Identify the indicators you will use to determine program success.
6. Specify the control mechanisms you will employ as necessary.
7. Rough out an appropriate design using Xs and Os (see Chapter 4) and tell why this strategy is preferable to others.
8. Be prepared to defend your recommendations to the superintendent.

A HYPOTHETICAL CASE: CUTBACK MANAGEMENT

Poro Valley, California, is a community of 300,000 persons located thirty-five miles southeast of Los Angeles. The city boasts a major steel manufacturing plant and several smaller manufacturing enterprises and acts as a bedroom community to a number of other industrial centers in southern California.

The community was devastated by the referendum known as Proposition 13, which rolled back state property taxes. The actual effects of the proposition on the cities of California was secondary because city revenues were not covered by the proposition. State contributions to city treasuries did drop drastically, thereby threatening many previously unquestioned programs.

For Poro Valley the most devastating impact of Proposition 13 was the ascendency of a Proposition 13 mentality among certain members of the community. Most notably, three of seven city council members co-chaired the local committee for passage of Proposition 13.

Poro Valley has a council-manager form of government. The city manager, who has occupied the position for fifteen years, enjoys adequate working relations with the council despite several major disagreements over funding for nonessential services. (A nonessential service, from the point of view of council hard-liners, is anything not having to do directly with police, fire, garbage, sewer and water services, or city maintenance.) As a result of Proposition 13, the manager has called together department heads and staff for the purposes of forming a cutback management task force. In all probability, there will be a cut of up to 15 percent in programs considered by the council as nonessential. Programs that do not directly service the public are to be the worst hit. The manager has decided to utilize two decision-making models. The first calls for across-the-board cuts in all programs, with extreme cuts in nonessential services. The second calls for elimination of some nonessential services so that others might be saved. The manager has instructed the task force to develop both scenarios so that the most programs can be saved.

One program to be particularly scrutinized is the fifteen-year-old tuition assistance program that helps city employees take courses that are applicable to their jobs. The courses may be vocational, managerial, or professional, and the program will pay for any expense that can be applied to an employee's job, whether the employee applies the course toward an academic degree or not. This program has come under direct attack by council militants as a giant bureaucratic boondoggle that does not service the needs of the community. They claim to have proof that tuition assistance has resulted in a building full of public employees who cheat the city out of a full day's work. Public employees, especially managers, are said to waste their days preparing for courses and cheat the city by using city clerical services for preparing of term papers. The militants have called for the firing of any employee so abusing the program.

The city manager and the personnel director both like the tuition assistance program. They believe that the benefits of the program far outweigh any minor instances of program abuse. A cursory check of the records revealed that less than 40 percent of the city's employees participated in the program over its fifteen-year existence.

The city manager is in a dilemma. He is convinced that the program is useful and rewarding for employees and the city, and he favors this program for nostalgic reasons: Tuition assistance was his first managerial innovation when he came to Poro Valley. The benefits of the program, however, must be weighed against those of other programs, which may be equally valued by city employees. Tuition assistance must therefore be considered in the context of the two decision models. Should it be maintained as is or eliminated, or should it be part of an across-the-board reduction in nonessential services.

In the cutback decision framework, there are three other programs against which tuition assistance must be weighed: employee parking benefits, the level of city contributions to employee retirement and insurance benefits, and the payment of employee membership dues in professional organizations. These programs are also the subject of controversy.

Parking in Poro Valley has not always been a problem. Fifteen years ago the city boasted wide avenues with diagonal parking spaces, but the population has since doubled. Under the manager's leadership, Poro Valley sought and received federal urban renewal funds to rejuvenate the inner city area. The renovation included a new government complex and two shopping centers designed to facilitate downtown shopping by city commuters living in outlying housing areas. As part of the renovation, downtown streets were all converted to one-way traffic, and street parking was eliminated in the vicinity of city hall except for fifty-cent-an-hour meter parking on two adjacent streets. City employees were given free parking privileges in the new municipal garage across the street from the city hall. These benefits represent a substantial savings to employees working in the building because parking in private facilities currently costs forty dollars a month.

All employees are not affected by the parking benefits. The police and fire department have parking facilities at the various precinct and fire substations. City sanitation workers and sewage treatment plant workers enjoy ample free parking in residential areas near those facilities. City street crews work out of the city garage, which is located at the southern edge of the city, where parking also is ample.

Insurance benefits and retirement packages for city employees have traditionally been excellent, with the city providing an 80-20 share of the cost of insurance and a 70-30 retirement package. The problems of cuts in these areas is compounded by the fact that Poro Valley public employees are completely unionized. The city conducts negotiations with the Fraternal Order of Police and the International Firefighters Association, which represent the city's uniformed personnel. The American Federation of State, County, and

Municipal Employees represents the clerical personnel, employees at the sewage plant, and street department workers. Sanitation workers are represented by the International Brotherhood of Teamsters. All four unions have threatened job actions if the city does not continue to provide insurance and retirement benefits. In fact, the teamsters and firefighters have proposed that the city assume a 100 percent share of the insurance benefit.

Payment of professional memberships is a program that spun off the contract the city manager negotiated with the city. Initially, his contract called for payment into a special annuity program for city managers and payment for the manager's membership in the International City Managers Association. As president of the regional chapter, the manager began to push for membership in the organization for all assistant managers and department heads. As an incentive, the manager secured a council commitment to pay for these memberships as well as his own. The program has since been expanded to cover all professional memberships, excluding labor unions only.

These nonessential services are not without support on the city council. Two members of the council are strong supporters of the manager. One, a dentist, was instrumental in the initial hiring of the manager. The other is a progressive business manager of a large electronics assembly plant. The remaining two members of the council generally tend to vote in support of the manager. Both have privately expressed their willingness to go along with anything the manager thinks necessary to sustain current service levels and balance the budget–insofar as possible. Publicly, however, they are reluctant to enter into open confrontation with the Proposition 13 militants.

The manager has instructed the task force to conduct full-scale evaluations of nonessential services. With regard to the tuition assistance program, his instructions were for the budget director, the personnel officer, and two members of the planning staff to develop an evaluation strategy to address the issues above.

Develop an evaluation strategy to address the above issues. What kinds of data that would facilitate an evaluation are available? What kinds of additional data must be generated in order to evaluate the program? What evaluation design could facilitate realistic comparisons between program participants and nonparticipants? Who are the probable consumers of the data? What format will best serve the needs of the evaluation consumers?

The evaluation team must determine whether the charges of the council militants are true. It must also determine whether there are appreciable benefits from the program to the city which can be used to justify the program's continuation. If employees had to choose, how would the tuition assistance program stack up against other benefits? Are there any differences between employees who have currently or formerly participated in the program and employees who have not participated? Begin your efforts by applying the outcome evaluation model on page 112 to the tuition assistance program.

FOR FURTHER READING

Dolbeare, Kenneth M., ed. *Public policy evaluation.* Vol. 2 of Sage Yearbooks in Politics and Public Policy. Beverly Hills, Calif.: Sage Publications, 1975.

Hatry, Harry P., Winnie, Richard E., and Risk, Donald M. *Practical program evaluation for state and local governments.* 2nd ed. Washington, D.C.: Urban Institute Press, 1981.

Rossi, Peter H., and Williams, Walter, eds. *Evaluating social programs: Theory, practice, and politics.* New York: Academic Press, 1972.

Rutman, Leonard, ed. *Evaluation research methods: A basic guide.* Beverly Hills, Calif.: Sage Publications, 1977.

Rutman, Leonard. *Planning useful evaluations: Evaluability assessment.* Beverly Hills, Calif.: Sage Publications, 1980.

Suchman, E. A. *Evaluative research: Principles and practice in public service and social action programs.* New York: Russell Sage Foundation, 1967.

Weiss, Carol H. *Evaluation research: Methods of assessing program effectiveness.* Englewood Cliffs, N.J.: Prentice-Hall, 1972.

Weiss, Carol H., ed. *Evaluating action programs: Readings in social action and education.* Boston: Allyn & Bacon, 1972.

Chapter 6

THE ART AND METHOD OF PROCESS EVALUATION

Process evaluations fill a gap left by goal-oriented outcome evaluations. In the words of one practicing administrator, "We are interested not so much in whether *X* causes *Y* as in the question if *Y* is not happening, what is wrong with *X*?" Although outcome evaluations reach valid conclusions that are important to top decision makers, outcome measures may not help line managers correct program deficiencies. The problem-solving focus distinguishes process evaluations from outcome-focused efforts. This chapter begins with a discussion of how process approaches differ from outcome evaluations in intent and methodology. The steps of a process evaluation and the methodologies employed are then presented using an education example.

THE PROCESS-OUTCOME DICHOTOMY

Michael Patton contends that the first step in an evaluation is to locate somebody who cares about the findings.[1] The evaluation must be targeted at those who will use the results. Outcome measures are excellent for legislators, account-

[1] Michael Quinn Patton, *Utilization focused evaluation* (Beverly Hills, Calif.: Sage Publications, 1978).

ing authorities (the Government Accounting Office, the Office of Management and Budget, and the like), and top administrators who make broad-based policy decisions. Evaluations commissioned by line managers are often carried out in the same outcome-oriented manner, but outcome approaches are of little value to program managers who want to know why the program is or is not working and what sort of program adaptations are appropriate.

A central issue in the debate between the outcome and process schools of evaluation is the defining of goals. Outcome evaluators believe that defined, measurable goals are prerequisite to evaluation. Process evaluators argue that goals cannot be firm because organizations are constantly evolving in response to changing conditions in the environment.

From the process standpoint, the important point is that programs do not begin at zero. Most programs are well established and ongoing, and theoretical and program goals were established in the near or distant past. Strategies for implementing the goals will have been adapted, new missions will have been tacked on to existing organization structures, the program staff will have long since ceased to occupy itself with questions of mission.

The skills of the process evaluator also are different. The outcome-oriented evaluator focuses on program outputs that are assessed against clearly defined program goals. The outcome evaluator's tools are knowledge of research design, sophistication in data gathering and analysis, and report writing. In process evaluation the consultant may be retained to assess program delivery systems, so the evaluator should also know something about the theory and practice of public program management and about the program under study.

This is not to suggest that process-oriented evaluators are not concerned with goals. Process consultants may also be retained to help reach a consensus on program goals. In this case, the process consultant might take the organization through a goal-defining process like that described in Chapter 5. The goal-consensus-building activities are not the exclusive stock-in-trade of the outcome evaluation specialist. In fact, concern for the *process* whereby goal consensus is reached is most typical of process consultants.

The outcome evaluator uses the defined goals to develop systematic measures of program outputs. The process consultant might use the goals to analyze the organization of agency activities, or to assess the allocation of organization resources among various activities, or to assist in the redesign of delivery systems in line with the goals.

The differences between the process and outcome approaches go beyond goal definition and include the research designs employed. Research designs appropriate to outcome evaluations were discussed in Chapter 4. We now turn to the methodologies appropriate to process evaluations.

APPROACHES TO PROCESS EVALUATION

The activities of the process evaluator may vary depending on the nature of the problem, the preferences of program staffs, the timing of the evaluation, and the amount of resources available for conducting the evaluation. Generally, however, the process evaluation model can be described as occurring in four phases, as shown in Table 6-1.

The first phase of a process evaluation–problem identification–is the most crucial element. Of course, evaluators may be asked to limit their activities to the identification of problems. In this phase the evaluator may use a range of techniques ranging from interviews with the agency director and program staffs to sophisticated research designs involving survey research and measures or program outputs. The second phase involves developing solutions to the problems identified in the first phase. These solutions may involve modest redistributions of resources among organization units or a complete reordering of agency activities. The third phase involves implementing the solution. In some cases this phase is just a passage of time between the decision to do something differently and the collection of data to determine whether the solution worked. When the solution involves major changes, it may be necessary to devise a change management structure to ensure that the solutions are not lost in the rush of day-to-day agency activities. The final phase of the process approach involves collection of data to determine whether the solutions were implemented as intended and whether the solutions produced the desired effects.

TABLE 6-1 The four phases of process evaluation

Phase I: Problem identification	***Phase II: Solution development***
a. The process consultant meets with program officials and engages in a series of problem identification activities. b. The consultant presents identified problems to program staff.	Program officials and the consultant select a course of action to resolve agency problems.
Phase III: Implementation	***Phase IV: Feedback evaluation***
a. The solutions are put into operation, with specific individuals taking responsibility for various components of the strategy for change. b. Management control systems are put in place to see that agreed-on changes are scheduled and carried out.	The consultant and/or program staff engage in systematic assessments of the impacts of the changes on the organization and program implementation.

The American public school system is a good case in point here. At the theoretical level, American society has established basic education at public expense as the right of all citizens. Equipping American youth with basic reading and math skills are program goals of the education system. Because these theoretical and program goals are taken as given by educators, the evaluator would not need to spend a great deal of time in goal definition activities before undertaking either an outcome evaluation or a process evaluation of a school district's reading and math programs.

To begin the discussion of process methodologies, suppose that a school administrator learned that the performance of district children on national achievement tests had dropped dramatically the previous year. Suppose further that past performance in the district had been constantly at or near the national average. Finally, suppose that the administrator had a feeling that the performance drop stemmed from an eighteen-month-old busing program that had affected teacher morale, caused anger in the community, and was thought to be the cause of discipline problems in the classroom.

Notice that the administrator's concern follows directly from the report of output measures of organization performance–reading and math scores. A prudent outcome-oriented evaluator would not wish to draw conclusions based on a onetime drop in student performance unless the drop was statistically well below previous trends. The evaluator would wish to collect more data over several years before concluding that a less than significant performance drop represented a downward trend. Unfortunately, the administrator can ill afford to wait for several subsequent data collections before acting. The administrator more likely would act immediately to assure that the drop was a onetime phenomenon.

PROBLEM IDENTIFICATION

The first task of the process evaluator in our example would be to ascertain whether the administrator's perceptions of the problem were correct. Several approaches are available and appropriate, including surveys (written or telephone) of teachers and parents in the district. The evaluator could also meet with teachers and parents in informal group discussions about district problems and the solutions the groups might suggest. Each approach has advantages and disadvantages.

SURVEYS

Formal surveys have several advantages. First, a written survey can ensure that a representative sampling of organization and community sentiments are measured through random sampling. Second, the data collected from each respondent can be addressed to a specific problem or problems by careful construction of the research questionnaire. Unfortunately, one cannot be absolutely certain that the questionnaire addresses the correct problem, but an open-ended set of items in which respondents are invited to clarify their questionnaire responses and express other issues of concern to them can help alleviate this disadvantage. The drawbacks of the survey approach are that surveys are expensive and time consuming. The decision to undertake a survey research project should therefore depend on the level of information certainty the organization needs and the resources that are available for the project.

GROUP INTERVIEWS

Group interviews can be carried out inexpensively and within a short time frame. The evaluator in our case would visit various schools in the district and meet with teachers and administrators to determine what they perceive the problems to be and what they think the solutions are. The evaluator can also meet with parents for the same purpose. Parent meetings could be arranged to coincide with regular parent-teacher or school meetings, or individual invitations to meet with the evaluator could be dispatched.

The Squeaking Wheel Syndrome. With the group meeting approach, the evaluator can make qualitative assessments of the nature and intensity of community and organization sentiments. Unfortunately, the consultant has no way of determining the extent to which sentiments expressed by parents who come to the meetings are representative of the community as a whole. For example, an evaluator might meet with a group of parents who were quite vocal in articulating their views on the busing program and its impacts on academic performance, but he or she has no way of determining whether the bellicose participants in the meeting represent widespread community sentiment or the ability of a well-organized minority to persuade its members to attend. We call this the squeaking wheel syndrome, which signifies that persons with intense feelings will attempt to impose their perceptions on others. The phenomenon may occur whenever information is sought in group formats. Meetings with teachers and administrators as well as with parents may be dominated by the bellicose or the articulate. Even in the absence of sharp divisions or extreme sentiments, the squeaking wheel phenomenon may occur as a result of a group's tendency to defer to its more vocal members. Unfortunately, these expressions may be an exercise in ego gratification rather than articulations of widespread organization or community sentiment.

The skills of the evaluator as a group discussion leader can help control the squeaking wheel syndrome. These skills cannot be used in a large public meeting and it may not be practical to provide the public small-group access to the consultant, but small group meetings with the staff are possible and desirable (the ideal group size is between eight and ten people). Given a manageable group size, the evaluator can guide the discussion, bring the group back to the problem at hand, and generally draw out inputs from the more reticent members.

The Despot/Suppression Syndrome. Group meetings with members of the organization's staff may produce an additional threat to the evaluator's effort to gain correct information: the despot/suppression syndrome. This syndrome is possible because administrative personnel exercise power over their subordinates, and therefore may suppress the expression of opinions contrary to their own. The process is usually subtle and is characterized by some combination of the following.

Prior to the evaluator's visit, the administrator holds a staff meeting and emphasizes the importance of presenting a united front and/or not airing dirty laundry in public. The administrator might also suggest that the staff confine themselves to responding to the evaluator's questions without volunteering additional information on "extraneous matters." In the meeting with the evaluator, the administrator also may engage in a variation on the squeaking wheel syndrome by focusing the discussion on a specific problem and taking issue with anybody who seeks to broaden the scope of the discussion, including the evaluator. Another tactic is to offer "clarifications" on every point made by the staff in order to portray the problem in the most favorable light; after several such clarifications, the evaluator will experience difficulty eliciting staff discussions.

The best way to control for the despot/suppression syndrome is to meet separately with staff and administrative personnel, but even separate meetings cannot rule out the impact of staff meetings that precede the evaluator's visit. An added benefit of separate meetings is that variations in perspectives are more likely to emerge under such circumstances. An administrator who meets with the consultant and the staff together may hold back to avoid dominating the discussion, in which case the evaluator receives only a staff perspective. Or the administrator may take an active part in the discussion to offer a broader perspective, in which case the staff may hold back in the belief that the evaluator is seeking only the broader perspective.

A final problem with group interviews involves consistency of outputs. When these meetings are carried out in an unstructured, open-ended fashion the community and staff can express whatever concerns them, but these concerns may not be related to the problem at hand. In our example, the evaluator wishes to determine whether the decline in reading and math scores is related to the busing program. A prudent evaluator wants to maintain a relaxed atmosphere in order to permit all problems to emerge. Unless the meeting agendas are structured, however, the discussion may degenerate into a debate over extraneous issues such as district salary and leave policies or inadequate audiovisual equipment or the formula for assigning lunchroom duties. By structuring the discussion, the evaluator can gain specific information regarding reading and math programs, the impacts of busing, classroom atmosphere, and teacher morale, and then open the discussion to other matters. Setting aside a portion of the meeting for unstructured input serves the same purpose as the open-ended section of the survey research questionnaire: both may produce information on problems not previously considered by the evaluator or the administrator.

SUPPLEMENTAL PROCEDURES

The survey research and group interview approaches provide the evaluator with staff and community perceptions of problems. Because these perceptions frequently do not tell the entire story, however, the evaluator may wish to engage in supplemental analysis before submitting recommendations. For example, the evaluator might use the participant-observer approach to get a firsthand view of discipline problems in the district. In addition, the evaluator should always undertake a review of organization structures and operating procedures to determine whether recent changes could be the cause of the problem at hand. We call this procedure a managerial audit. Both these supplemental procedures would enhance the evaluator's analysis in our example.

The Participant-Observer Approach. In the current example, the evaluator could collect information on classroom discipline from the teacher surveys or group interviews. Supplemental information from school discipline records also could be gathered. Classroom visits, however, could give an evaluator a sense of district problems which cannot be gained from statistical analysis or interviews.

The participant-observer approach has roots in the research traditions of cultural anthropology, where it involves visiting a subject culture and studying how various functions common to all cultures are performed. The methodology has been popular among those who evaluate education programs because it allows the researcher to make qualitative as well as quantitative assessments of program content. In our example, the evaluator could visit district schools and qualitatively assess classroom atmosphere and observe the children at play, thereby determining which students or groups of students were causing problems. The evaluator could also determine whether teachers in a given school treated children differently on the basis of race. The latter is of particular importance because solutions that are appropriate when teachers are at the root of the problem are different from solutions that are appropriate if interracial tension between student groups is the problem.[2]

---

[2] For a readable account of how teachers' attitudes can cause students to behave in a manner consistent with those attitudes, see William Ryan, *Blaming the victim* (New York: Vintage Books, 1976).

An additional benefit of the participant-observer approach is that the staff is more likely to accept an analysis when the evaluator has had an opportunity to view the problem from an operational perspective. In the current example, staff familiarity with the methodology and its appropriateness to education programs will make acceptance of the evaluator's analysis likely.

The Managerial Audit. In order to effectively assess organization structures and operating procedures, an evaluator should be well acquainted with organization theory and the practice of management. In addition, the evaluator should be sensitive to recent changes in program procedures or personnel policies. Suppose the evaluator learned that two years earlier the district had begun both a controversial reading program and an open-classroom format in several schools and that, in addition, there had been a 40 percent turnover in teachers over the past three years because of a liberalized retirement policy and the fact that senior teachers had been leaving the district for better paying jobs in nearby towns. All these factors could contribute to a decline in performance scores in the district. Such data might not be elicited by the survey or group interview approaches, especially if the dominant belief was that performance declines were the direct result of the busing program.

Managerial audits are particularly beneficial to the process evaluator for two reasons: First, they are relatively inexpensive to carry out because they mainly involve discussions between the evaluator and administrative personnel and reviews of operating procedures.

Second, and more important, is the fact that an analysis of organization operating procedures is a prerequisite to making implementable recommendations. Suppose, for example, that teacher conduct toward students was an identified problem but the district had a collective bargaining agreement with its teachers. The evaluator and the administrator would need to involve the union leadership in the design of a program to resocialize teachers. Failure to work with the union could result in personnel grievances being filed and possible court litigation involving the district and individual members of the bargaining unit. In other words, the district's collective-bargaining agreement served as a constraint on effective change strategies.

PRESENTING THE FINDINGS

Upon completion of the problem identification phase, the task becomes one of presenting the findings to the administrator. To illustrate this process, we assume (1) that the evaluator had employed the group interview approach, the participant-observer method, and a managerial audit; (2) that the evaluator concluded that teacher turnover and program changes, not the busing program, were the principal causes of the drop in performance scores; (3) that the evaluator had witnessed differential treatment of students on the basis of race by a number of district teachers; and (4) that the group interview process indicated that antibusing sentiment in the district was high and that teachers as well as parents believed busing to be at the root of the declines in performance scores.

Identifying the problems and having solutions that correctly address them are not enough. The evaluator must convince the consumers (the administrator, the staff, the community) that the analysis is correct. In our case, the administrator might not want to believe that the reading program and/or the open-classroom format contributed to the problem, especially if the administrator was instrumental in installing the two programs. Similarly, teacher turnover rates may have resulted in part from the administrator's hard-line bargaining tactics in recent years. In such a case, the administrator must be persuaded to accept the evaluator's analysis as correct.

The evaluator's principal asset in the persuasion process is the ability to present information in a clear and concise manner that is not threatening to the audience. In the example, the evaluator might begin by reporting the group interview findings that largely support the administrator's perceptions that busing was the principal problem. The fact that the consensus in the district is that busing is a problem makes action necessary, even though it is not the principal cause of the decline in scores. Next, the evaluator could present the analysis of differential treatment of children by teachers on the basis of race. Only after illustrating that busing and discipline problems were more complex than previously thought should the evaluator proceed to a discussion of the reading program, open-classroom formats, and teacher turnover rates as causal factors in the declining scores.

When an evaluator is presenting a controversial analysis, it is well to enter the discussions with as much supporting evidence as possible. In the current example, the relationship of score declines to factors other than busing was based on the expert analysis of the evaluator. Whether or not an evaluator's expert analysis is accepted depends on how much credibility the evaluator has with the audience and how persuasive the presentation is. If the administrator remained unconvinced, it might be necessary to reiterate and once again undertake the problem identification process by gathering empirical data on performance scores controlled for the conditions in question. For example, performance scores could be reanalyzed with controls for teacher seniority in order to test whether teacher experience was a factor in student performance. The data also could be reanalyzed on the basis of whether students were in open or traditional classrooms. Such secondary analyses assume that student performances can be identified by classroom format and teacher experience. To ensure that the analysis measured the impacts of seniority rather than the student's ability, the evaluator also would have to be certain that teacher assignments to classes of more gifted children were not made on the basis of seniority.

Because the controversial reading program was implemented districtwide, testing its effect would require an experiment in which students would be randomly assigned to the program or to a more traditional approach (see the discussion of experimental methodology in Chapter 4). Such an experiment would provide concrete data on the program. The experiment, moreover, could be carried out on a limited basis, thus avoiding the expense of districtwide application.

It is also possible that the evaluator can persuade the administrator on the basis of his or her expertise and on the logic of the presentation. Whether by empirical data or by persuasion, convincing the administrator of the wisdom of the evaluator's analysis is prerequisite to developing solutions to the various problems in the organization.

SOLUTION DEVELOPMENT

Once the administrator is convinced that performance declines are related to factors other than busing, the next step would be to work out a set of solutions. Whenever possible, these solutions should be devised in close consultation with program personnel, including the administrator and his or her assistants and with as much staff input as necessary and affordable.

The goal of the consultative problem-solving mode is to generate solutions that are workable, acceptable to line personnel, and affordable. In the current example, there is strong administrative support for the reading program and the open-classroom approach. Converting to more traditional reading programs and discontinuing the open-classroom approach would require considerable expenditures for purchasing teaching materials and for orienting teachers to the replacement program, not to mention the cost of reconverting the open classrooms.

The possibility also exists that the decline in performance scores is the result of the transition to the open-classroom format and teacher adaptation to the controversial reading program. If this were the case, scores could be expected to normalize and even improve over time. The possibility also exists, however, that teachers either do not fully understand the methodologies associated with open classrooms or the reading program, or that teachers might be opposed to the approaches and therefore be resisting correct implementation. In either case, program modifications would be preferable to wholesale program replacement.

The district could determine whether the problems were transitional by allowing them to continue for a time to see if performance scores improved.

During the wait the district may wish to employ the participant-observer approach to see whether the programs were being implemented as intended. If they were not, a training program could correct perceived difficulties, whether they be caused by misunderstanding or by misapplying program procedures. If the adaptations did not result in improved scores within a reasonable period of time, the district might wish to consider discontinuing them.

The problem of differential treatment of students by teachers on the basis of race would remain whether or not curricula and classroom formats were altered. The problem could be addressed by a program in human relations training that would be aimed at making teachers aware of their personal belief systems and the impact of those beliefs on the way they treat students. The program should not be expected to change basic beliefs, but on completion of the training, teachers could be enlisted to design systems whereby teachers could mutually assist one another in carrying out a districtwide program to bring about equality of treatment.

Resolving the problem of teacher turnover rates would require a change in district salary policies and the strategies the district employs in the collective-bargaining process. First, the district should adjust its salary schedule to bring it into line with prevailing rates in the region. Second, the bargaining strategies of the management team should temper the desire to counter any and all union demands by recognizing the district's need for experienced teachers and the costs to the district of high turnover rates. Money saved at the bargaining table can be lost in the training of new teachers and losses in the quality of instruction in the district. Once high turnover rates are stemmed, they should become self-reducing over time as younger teachers become more proficient. The district might wish to accelerate this process by instituting in-house training in which senior teachers shared their experiences with younger colleagues.

All these solutions are aimed at the real causes of the district's declining reading and math scores, but the problem of busing also must be dealt with so that the other solutions have a chance to work. The misperceptions regarding busing could be countered by a public education program designed to inform the community of the evaluation findings, the solutions being undertaken, and the results being sought. Such a program, however, will provide little solace to those who view busing as a problem in and of itself. If antibusing sentiments were widespread in the district, and if the busing had been court ordered, the only recourse for the district would be to manage the impacts of the busing. To accomplish this, the evaluator might recommend that the district undertake a study of the current busing program in light of the racial distribution of students in the district. Using computer-assisted analysis, the court-ordered busing could be implemented by moving a minimum of children a minimum distance.

In sum, solution development in process evaluations should be worked out in consultation with program staff to maximize staff acceptance and cooperation in implementation. As far as possible, solutions should emphasize adaptations in current programs rather than wholesale program alterations or discontinuance. In addition to program adaptations, the solutions in our example involved changes in administrative procedures (such as salary schedules and negotiation strategies), training for teachers (including human relations training and skills training for younger teachers), a computer-based redesign of the busing program, and a public information program.

What was initially believed to be a problem of busing impacts on performance scores has been identified as a much more complex problem with several root causes. The proposed solutions are complex and potentially disruptive of routine activities in the district. Conversely, the danger exists that the well-intended solutions could be lost in the press of routine education processes. Minimizing disruptions and ensuring success requires a carefully designed strategy to manage implementation of the solutions.

IMPLEMENTATION

The management of program adaptations is the single most overlooked phase of the process approach. Regardless of whether one is a line manager or a program evaluator, one must recognize that change is frequently resisted and nearly always disruptive. Knowing what the problems are and knowing how one wishes to do things differently are not enough. One also must design a change management system to minimize the potential for staff resistance and the disruptive effects of the proposed changes on daily operations.

There are two components to any planned change process: a clear agenda for change and a system for managing implementation of the agenda. The adaptations to be carried out have already been described. The focus here is on the management of the proposed changes. The two essential elements to a successful change management strategy are (1) the unequivocal support of top management and (2) a clear delineation of responsibility for carrying out the change. The support of top management can be expressed in memos to the staff and orally during staff meetings. The administrator can further show commitment by scheduling a series of meetings in which the responsible staff members would be required to report their progress to the administrator.

The success of proposed changes is also affected by the personnel who carry them out. There are three ways to guarantee failure. The first is to announce the changes along with the expectation that everybody in an organization will assume responsibility for carrying them out. The second way to ensure failure is to assign responsibility for the program to persons who lack the skills to carry them out. The third way is to assign the responsibility to competent persons with the stipulation that changes be carried out in addition to regular duties.

THE TASK FORCE APPROACH

Ideally the administrator would assemble a task force of expert personnel to implement the changes. Task force leadership should be delegated to the administrator's most competent deputy or be assumed personally by the administrator. Membership in the task force in our example might include the district's reading and math specialists, the program evaluator, the director of employee development, the district's personnel officer, a representative from automatic data processing, and representatives of the teachers' union, the school board, and the parent-teacher association. Inclusion of the various district specialists and the evaluator would ensure that the task force had the expertise necessary to carry out the proposed changes. The presence of the board members, parents, and teachers' union representatives would help ensure cooperation of key groups in the implementation of the changes.

In addition to assuming overall responsibility for the changes, the task force would act as a sounding board for ideas and a coordinating body to ensure that various components of the planned changes were carried out on schedule. The task force would also be responsible for preparing the final report to the administrator and school board.

WORK GROUPS

Actual implementation of the various changes would be carried out by work groups composed of task force members and relevant program staffs. The appropriate number of work groups varies with the complexity of the change program. In the current example, three work groups could carry out the changes. The groups would include a curriculum adaptation work group, a group responsible for human relations training for teachers, and a management adaptation work group.

Curriculum Adaptation Work Group. The curriculum adaptation work group would be responsible for carrying out participant-observer visits to the classrooms to determine whether the reading program and the open-classroom format were being correctly applied. The group would also be responsible for the design and implementation of any necessary training to reeducate teachers. Because of the centrality of this group to a successful adaptation, the task force head may wish to chair this work group personally (at least the task force leader should meet regularly with the group). The other members of the group would include the district's curriculum specialists (reading and math), one or more principals of schools using the open-classroom format, the district's training officer, and the evaluator. The district's training officer must be a member of the group in order to schedule and implement any prescribed training. The evaluator's presence is necessary to ensure that the adaptations undertaken by the group are accounted for in the postadaptation evaluation design.

Human Relations Work Group. The human relations work group would be responsible for making teachers aware of how their personal values and attitudes toward persons of other racial and ethnic groups can affect their classroom performance. Most school districts do not have the staff resources to carry out human relations training. When such training is necessary, organizations generally contract for the services of human relations experts. Whether the training is contracted or conducted in-house, the district must provide the necessary logistical support, which would include provision of training facilities, scheduling the release of teachers to undergo the training, and the like.

A logical follow-up to the human relations workshops would be teacher support groups in each school which would be chaired by a designated human relations coordinator selected from among the ranks of classroom teachers. These groups would meet regularly with teachers to discuss problems related to applying the lessons learned in the workshops to actual classroom situations. In time, these groups might no longer be needed. The support groups could become a permanent feature in the district if the discussions that began with a human relations focus were to evolve into a forum where teachers could share mutual problems and frustrations related to teaching in general.

The human relations work group would be composed of the district's director of employee development, who would chair the group; a human relations coordinator for each school in the district, selected from among the ranks of practicing teachers; a human relations consultant, who would conduct the workshops; the program evaluator, who would serve as a resource and design the measure to evaluate the outcomes of the group's efforts; and a representative of the teacher union, to ensure that teachers' rights were protected and to make it less likely that teachers would resist participation.

Management Operations Work Group. The management operations work group would be responsible for carrying out the administrative components of the change design, including conducting a wage and salary survey to determine whether salaries should be adjusted. The group would also be responsible for studying the district's busing formula and making recommendations for a busing plan based on residence and voluntary participation. Because this study would require a sophisticated computer-assisted analysis, an external consultant might be necessary. Finally, this work group would review the district's collective-bargaining strategies with an eye to tempering management negotiation zeal with a concern for the staffing needs of the district and the cost of high turnover rates.

Because this work group's responsibilities are administrative, its composition could be limited to administrative personnel. The administrator or a trusted assistant should chair the group. The wage and salary study could be carried out by the personnel division or be delegated to the evaluator, who would also be a member of this work group. In addition, the group would include a representative of the district's automatic data-processing unit. The collective-bargaining adaptation would make it necessary to include a member of the district's bargaining team and a representative of the personnel division.

The composition of the various work groups reflects an assumption that the hypothetical school district was large enough to have the various support divisions (for example, reading and math specialists, personnel specialists, automatic data processing capacity). In smaller districts, these functions may be carried out by a small number of all-purpose administrative aides and classroom teachers, and greater emphasis might be placed on external consulting services. Whether the district is large or small, however, it is important that responsibilities for change be clearly assigned to specific persons for implementation.

SCHEDULING TASKS AND SETTING UP A REPORTING SYSTEM

The final detail in the implementation phase of the process model is a schedule for completing the various tasks. The administrator can know whether things are proceeding as planned by scheduling regular progress report meetings with the leaders of the various work groups. These meetings would also provide group leaders with access to the administrator, so that the latter could assist in persuading resistant individuals or groups to cooperate with the change program. They also would allow participants to adapt to unforeseen obstacles, such as delays in the receipt of questionnaires, breakdowns in the data-processing system, or the press of other matters that required reassignment of personnel.

The amount of formality needed in the reporting system varies with how complex the proposed changes are and how much personal control the administrator or task force leader wishes to exercise over the process. In relatively simple change programs (such as our school example), the level of control might be limited to a reporting schedule that coincides with the expected completion dates of various components of the plan. For complex change programs, it may be worthwhile to develop a program plan utilizing the techniques of PERT/CPM, described in Chapter 2. In either case, a formal reporting system at the outset gives employees a clear picture of what is expected of them and serves to reinforce managerial commitment to the proposed changes.

A two-part accountability system would suffice for our education example. Part 1 is what Beckhard and Harris termed a responsibility matrix.[3] Part 2 is a

---

[3] Richard Beckhard and Ruben T. Harris, *Organizational transition: Managing complex change* (Reading, Mass.: Addison-Wesley, 1977).

simple project calendar containing a timetable of project components and due dates for their completion.

The Responsibility Matrix. Consultants use responsibility matrices to help organizations engaged in change strategies. The matrix is a written record of who is responsible for each planned change (see Figure 6-1). The names of individuals are recorded across the top of the matrix; the various components of the change strategy are listed on the left-hand margin. A responsibility matrix uses a four-letter code to delineate responsibilities. The symbol *A* stands for approval/veto authority. An *A* beneath an individual's name means that the person must Approve of, and may disapprove of, specific activities designed to effect changes. Normally, however, these people do not engage in wholesale redesigns of agreed-on program changes. Because of the authority an *A* designation represents, it is best to limit its designation to one or at most two members of the organization. Having two people with veto authority over one or another component of a planned change makes it difficult for those responsible for the project to perform their assigned functions. The project manager, who is designated by an *R*, for Responsibility, in the matrix, should not have to serve multiple masters.

The project manager is assisted by other organization actors, who are designated by an *S*, for Support, in the matrix. *S*'s are also responsible for a project's success, although they usually have other duties that are of a greater priority for them. The project manager is responsible for leadership, for coordinating work group activities, and for the timely completion of tasks in the group. A project

	District Superintendent	Chief Adm. Assistant	Employee Dev. Specialists	Personnel Officer	Director Data Processing	Principal School A	Principal School B	Principal School C	Principal School D	Teacher School A	Teacher School B	Teacher School C	Teacher School D	Representative Union	Dist. Math Specialist	Dist. Reading Specialist	Evaluation Consultant	Human Relations Trainer
Curriculum Adaptation Work Group	*A*	*R*	*S*	*I*		*S*	*S*	*I*	*I*					*I*	*S*	*S*	*S*	
Human Relations Work Group	*A*		*R*	*I*		*I*	*I*	*I*	*I*	*S*	*S*	*S*	*S*				*S*	*S*
Management Operations Work Group	*A*	*I*	*S*		*R*	*I*	*I*	*S*	*S*							*I*	*S*	

Code: *A = Authorization Veto*
R = Responsibility
S = Support
I = Information

FIGURE 6-1
Responsibility matrix for a hypothetical school district

manager may also be assigned to a work group on the basis of expertise rather than rank in the organization, and members of the group may be the project manager's peers or superiors in other organization matters. The person possessing the *A* designation, therefore, must be sufficiently high in the organization to compel cooperation from reluctant members of the work group. Without this type of accountability, persons on project teams can obstruct activities to which they are opposed or at least let support for the project slip in the press of day-to-day duties.

The final symbol in the matrix is an *I*, for Information, which means that the person in question needs to be kept abreast of work group activities. Such information may merely be a matter of organizational courtesy, but it is more likely that persons given the designation *I* need the information to anticipate what impact work group activities will have on their organization responsibilities.

The sample matrix in Figure 6-1 utilizes job titles rather than names of individuals across the top of the matrix. When dealing with real organizations, it is best to use the names of the participants as well as their titles. To simplify the illustration, it was assumed that there were only four schools in the district. A district with problems as large as those listed earlier (especially with a busing program) would have many more schools. We elected to reserve the approval/veto authority to the district superintendent. This is possible because of the presumed size of the organization and because it is a school district. Schools are organizationally decentralized and have flat hierarchies. In an organization with a stricter hierarchy or a greater diversity of mission, a wider distribution of the approval/veto authority would be necessary.

The Project Calendar. The second component of the control system is a project calendar that indicates the components of the planned change and the inclusive dates for completing various phases of the project. To illustrate the uses of a project calendar, assume that the district has allocated two years for completing all phases of the adaptation and for carrying through a follow-up evaluation. Figure 6-2 is a project calendar for changes in our imaginary district.

Across the top of the calendar in Figure 6-2 the months of the project are numbered sequentially from 1 to 24. Relevant dates in the school calendar also are indicated. The large *V* in the display indicates the current date. In our example, the *V* indicates March 15.

The project work groups and an evaluator's calendar are listed on the left. Each project is tracked by a bar that indicates how each project component is progressing. As a project progresses, the team leader and task force leader fill in the blank bar. Thus, a quick check of the calendar reveals that the Curriculum Adaptation (CA) and Human Relations (HR) work groups are behind schedule, while the wage and salary (WS), busing revision (B), and collective-bargaining projects (CB) are proceeding according to plan. The letter/number codes above the bars in Figure 6-2 indicate the various phases of the project, which are as follows:

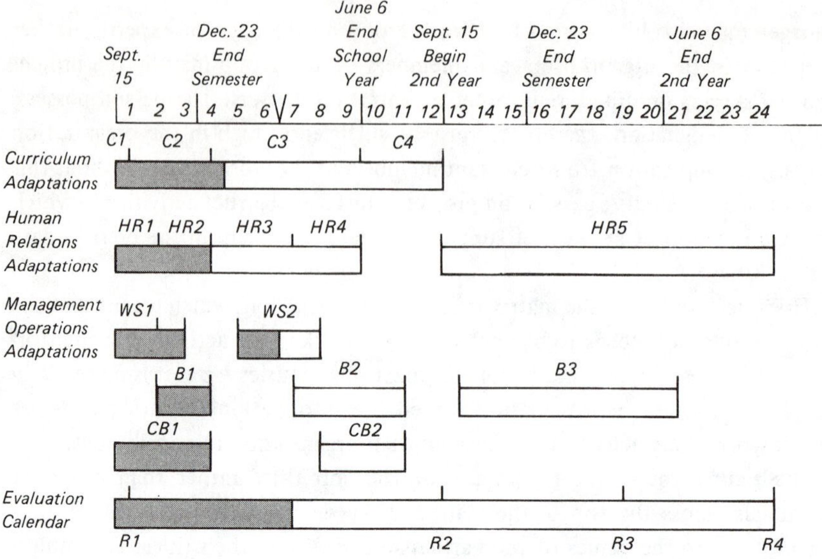

FIGURE 6-2
Project calendar for changes in a hypothetical school district

Curriculum Adaptations

C1: The reading and math specialists undertake classroom visits to identify incorrect applications of the district's reading and math learning programs.

C2: The specialists and the training coordinator conduct courses in how to implement the programs properly.

C3: The remainder of the school year is used to implement the revised methods.

C4: The methods are taught to teachers coming to the district for the first time.

Human Relations Work Group

HR1: Indicates the period in which the consultant is hired and the training program for teachers is designed.

HR2: The HR leaders for each school are selected and trained.

HR3: The remainder of teachers undergo human relations training.

HR4: Involves the application of what the teachers have learned in the classroom and playground setting. They are supported in their efforts by regular meetings with the HR group in their school.

HR5: The HR groups are converted into quality circles in which the teachers share experiences and problems to mutually support each other's efforts to become more effective teachers.

Management Operations Adaptations

There are three simultaneous projects in the planned managerial changes: A wage and salary survey (WS), a busing study (B), and a collective-bargaining strategy study (CB), each of which has more than one phase.

WS1: Involves a survey of other districts to determine salary comparabilities with the district.

WS2: Involves submitting the findings of the study to the school board for consideration in order to anticipate changes in the budget that will be necessary prior to the upcoming contract negotiations.

B1: Allows for the letting of contracts for the study.

B2: The contractor conducts the study and presents the findings to the district by the end of the current academic year.

B3: Involves implementation of the revised program in the second year of planned adaptations.

CB1: Involves a review of the district's previous contracts and union demands as well as a review of alternative approaches to bargaining.

CB2: The revised strategy is presented to management for study in the context of the findings of the wage and salary survey.

Evaluation Calendar

The final bar in the illustration represents the schedule of reports that are due from the evaluator.

R1: A report on the reading and performance scores of students in the current year compared with previous years, and a report of the findings of the reading and math skills specialists.

R2: A report of the findings of test scores at the end of the first year of the adaptations. This report might also contain recommendations based on unanticipated events that arose during the execution of the first year's adaptations. And the evaluator might be called on to survey teachers as to their perceptions of the human relations training they underwent and the effects of that training on their classroom performance.

R3: Another interim report on the successes of the curriculum adaptations, the human relations training groups, and the successes of the first semester of the new busing efforts.

R4: The consultant's final report at the end of the two sections, which includes empirical assessments of all phases of the program.

Keeping on Schedule. Keeping a project moving on schedule is the responsibility of each work group leader and the task force director. When slippages in the schedule occur, such as those in the curriculum adaptations and the human relations components of our example, prompt action is necessary. Administrations must decide whether the slippage is the result of poor cooperation with work group leaders on the part of program staff or of unrealistic expectations in the original plan. If the problem is one of cooperation, the responsibility matrix can help direct corrective measures. If the problem is poor planning, schedule revisions are in order.

The use of such managerial aids as program calendars and responsibility matrices cannot in and of themselves ensure the timely completion of program adaptations, but when combined with a real commitment to change on the part of management, these control aids can be extremely beneficial.

FEEDBACK EVALUATION

The final phase of the process approach involves gathering data on the impacts of the change program. The degree of sophistication necessary in this phase varies with the needs and preferences of the consumers, the skills and preferences of the evaluator, and more important, the nature of the changes enacted. In the case of human relations training of teachers, evaluation might involve visits to schools to see if training had the desired impact. The evaluator might also interview teachers individually or in small groups to determine their impressions of the training process and the impact of the training on teacher behavior in the classroom.

Feedback on the administrative aspects of the changes would involve a written report on the salary survey and the outcome of the computer-based busing analysis. Proposals for changes in the bargaining strategy of the district could be made directly to the administrator by staff members. The true measure of changes in the process, however, would be the contents of subsequent contracts and teacher turnover rates in ensuing years.

The most challenging aspect of the feedback phase of our example would be collection of data on student performance scores after the changes were implemented. The administrator might wish to look at the performance scores only in the aggregate to determine the net impact of the various changes. Or the consultant could design an evaluation that controls for teacher experience, open versus traditional classroom formats, the racial composition of student populations in the various schools, classroom ratios of bused students, and so on. The only way to determine whether the reading program and open-classroom format are being correctly implemented, however, is by continued direct observations. The more complex design would be desirable to ensure that each element of the change program has the desired effect. When the administrator has data on each element, he or she can make specific additional adaptations that would not have been possible based on an analysis that considers only student scores in the aggregate.

It is worth noting that the design possibilities are flexible because the evaluator was present when the changes were being designed and continued to be involved throughout the implementation phase. Such flexible designs are not possible when the evaluator is brought in after implementation.

OVERVIEW OF PROCESS EVALUATION

A four-phase approach to process methodology can be adapted to most organizations. In our example of a hypothetical school district, we showed how the evaluator must balance the application of empirical data-gathering skills with skills in communications and consensus building and a knowledge of organizations and their operation. Flexibility of design is essential for attacking the problems of an agency. A process evaluator might use group interviews, managerial audits, or the participant-observer approach, in addition to survey research and systematic data gathering. Each method has its appropriate applications and limitations—none is universally applicable. The art of evaluation is knowing which methodology is appropriate and applying it accordingly.

Finally, when one engages in process evaluation one places oneself at risk because the process evaluator is expected to recommend program adaptations as well as point out program malfunctions. The evaluator therefore assumes a degree of responsibility for program success as well as for evaluation accuracy. Because the evaluator shares ownership of program adaptations, he or she is more likely to actively assist in their implementation. Therein lies the principal difference between outcome evaluators and process evaluators. Outcome evaluators emphasize measurement accuracy, process evaluators emphasize problem solving.

Case Studies

A HYPOTHETICAL CASE: BALKINWALK, VIRGINIA

The city of Balkinwalk, Virginia, is a typical mid-Atlantic seaboard city of 750,000 which is growing quickly as a result of the migration away from the northeast. As with large cities in the state, Balkinwalk is responsible for the administration of social service programs as well as municipal services; in smaller jurisdictions, social services are county functions.

The new city manager has been in Balkinwalk for less than a year, during which time the majority of complaints coming into her office are about the social welfare department. Citizens complain that they are not treated well at the social service agency, that they are made to wait for long periods by insensitive clerical staffs, then told to come back the next day. The city manager has been passing such complaints along to the administrator of social services, but now she feels that she must take a personal hand to remedy the situation.

The current head of the social service agency, who has held that position for eighteen months, supervised the social workers for the previous five years. Before that, she was a practicing social worker in the Balkinwalk agency for eight years. The previous administrator of the social welfare department, like his predecessor, had been hired from outside the agency. The immediate predecessor stayed eleven months, then left for a better-paying job. His predecessor had remained in the job for just under two years.

The head of social services is not insensitive to the city manager's concerns. She too has sensed that something is wrong and that service could be better. From her perspective, the primary problems stem from the heads of various branches of social services who seem bent on frustrating each other and her.

The service delivery operation is made up of three units. The eligibility unit employs twenty workers and one supervisor, all of whom are paraprofessionals and whose job is to determine benefit eligibility. These workers have acquired a good deal of knowledge regarding policies and procedures. The administrator is aware that workers in this unit resent the low wages and benefits they receive. In addition, they have expressed scorn for social workers, whom the eligibility workers believe to be "a bunch of overpaid smart-alecky college kids." Eligibility workers also resent the hiring of what they perceive to be underqualified minorities for positions in their unit. Recently, these unhappy workers began an employees' association, even though there is no law in the state either permitting or forbidding collective bargaining.

The other two units are made up of social workers: the family counseling unit (twelve workers) and the individual case workers (eighteen workers). Each unit is headed by a "working supervisor." Social workers resent the amount of paperwork they must do as well as the caseloads they must manage. The large caseloads preclude the giving of individualized counseling to even the most needy clients, and the social workers believe that routine clerical tasks could be delegated to eligibility workers. The discontent among social workers has resulted in a high turnover rate, which some attribute to caseloads and others to the attraction of higher-paying jobs in nearby cities.

The personnel unit within the social welfare agency is responsible for recruiting and training personnel for the social service agency. In addition, it oversees and implements the social services component of the citywide affirmative action plan. The personnel director has held that post for less than nine months but has already incurred the anger and resentment of the program staff because she insists on vigorous enforcement of the affirmative action plan.

The accounting and payroll unit has two functions. The first entails preparing of the necessary forms so that the regional office can make payments to aid recipients. Until recently, the checks were dispensed locally, but the new regional procedure requires redesigning tasks and a new technology in the form of computerized records. This changeover requires the acquisition of new skills either by training current personnel or by recruiting new people from the outside. Accounting and payroll does not believe that the personnel department has been helpful in this regard.

The payroll division of the accounting and payroll department is also encountering its share of employee anger and resentment. The agency lost a recent employee appeal to the Civil Service Commission. The appeal was filed by a black female employee who had been fired for poor attendance and punctuality during her probationary period. The agency lost because it could not document the employee's failures. In reaction to this loss, the department installed a time clock, which has alienated all employees, including the social workers who frequently have to drive long distances to punch out at the appointed hour, causing further reductions in service to clients.

In consultation with the city manager, the agency head has decided to seek expert help to answer the following questions:

1. What is the level of service in our agency?
2. How do we stack up against other comparable programs?
3. Is the turnover rate inordinately high among social workers?
4. Do all clients share the sentiments of those who are complaining to the manager?
5. What can the agency do to extricate itself from the dilemma?

Assuming that the answer to some or all of the above questions turned out to be yes, design an evaluation project in the process mode to help the agency define and solve its problem. How would you proceed? From whom do you collect data? Do you merely interview staff or should clientele also be surveyed? What can be done to help the agency get back on the right track? How can the evaluator find someone who cares about the evaluation toward whom he or she can direct the findings?

A HYPOTHETICAL CASE: A CHANGE OF COMMAND

The Third Battalion, Forty-second Field Artillery, is a proud unit with a long and distinguished record of service to the nation in time of war. The peacetime mission of the unit is to engage in training in order to maintain the highest degree of readiness.

Organizationally, the battalion consists of three separate firing batteries, each of which is equipped with six 155 millimeter howitzers. In addition, the battalion has a headquarters unit and a service unit (the latter two are also called batteries). At full strength, the battalion consists of 210 personnel, including a complement of 30 commissioned officers and 70 noncommissioned officers.

The operational mission of the battalion is the conventional and nuclear fire support of the First Brigade, Fifty-second Mechanized Infantry Division. When engaged in support fire activities, each of the three batteries is assigned to support a specific maneuver battalion of the brigade. When in operation, the artillery battalion is dispersed, making coordination difficult. Thus, effective command requires a high level of training, mutual loyalty among the officers, and effective lines of communication–both electronic and interpersonal. Maintaining the level of training and loyalty is the responsibility of the headquarters and service batteries. Effective communication is a function of the commanding officer and the tone of communication among the officers.

For four years, command of the battalion was in the hands of Lieutenant Colonel Rock Smirnoff, who made it known at the outset that his goals were a vigorous program of personal fitness, a smooth-running command, and an uneventful final tour before his retirement. It is not surprising that Lieutenant Colonel Smirnoff delegated many command responsibilities to his executive officer, Major Jackson Beauregard Daniels. Major Daniels proved to be more than competent, and the performance of the battalion on such things as its efficiency rating, Army Training and Education Program, army-wide Skills Qualification Test (SQT), and Inspector General (IG) reports was consistently superior.

Eighteen months ago, however, Major Daniels received a transfer and was replaced by a young major named N. C. Cureley, who did his best but lacked the personal assertiveness of Major Daniels. Major Cureley refused to take responsibility for making command-level decisions, and both morale and performance declined until the retirement of Lieutenant Colonel Smirnoff three months ago.

Lieutenant Colonel Smirnoff's replacement was an upwardly mobile academy graduate with stars in his stars named I. M. Acomer. Lieutenant Colonel Acomer immediately assumed command of everything and responsibility for decisions down to the battery level and below. Discipline problems at the squad level were not beneath his concern. His organization goal (as stated at the first officer's call) was "to build the best battalion in the army." An inspector-general's report just received, however, indicated no improvement in morale or overall battalion performance. In fact, reported incidents of disciplinary problems were increasing. Realizing that something was seriously wrong, Lieutenant Colonel Acomer approached the base's organization effectiveness officer to get help diagnosing and solving the unit's problems. The officer took the commander and senior officers on a weekend retreat, during which they arrived at a set of definable problems and a set of solutions. The problems identified included:

1. Ineffective communications
2. Failure to delegate authority
3. Low morale
4. Discipline cases involving alcohol and drug abuse
5. Declining reenlistments
6. Declines in overall combat readiness
7. Equipment problems traceable to failure to accomplish scheduled maintenance.

The officers agreed to set up a program to correct the differences involving lines of communications, to develop a more cooperative mode of management involving delegation and trust, to begin a stepped-up program of training, to go by the book on scheduled maintenance and dealing with subordinates, and to commit themselves to assertive leadership by example and by an overall commitment to the organization.

Lieutenant Colonel Acomer has agreed to this program, but he wants the proposed changes to be monitored to see whether the desired effects are achieved. To his delight, the commander learns that three of his officers (two captains and a lieutenant) are enrolled in a masters of public administration program provided by a nearby university. Lieutenant Colonel Acomer calls in the junior officers and informs them that they are responsible for providing an objective evaluation of the battalion's performance. The evaluation report is due in twelve months, but an evaluation design is due on Acomer's desk in two weeks.

Assume you are the officer in charge of the evaluation. How will you proceed? Develop an evaluation strategy, then provide Lieutenant Colonel Acomer with hard data on the battalion's progress.

A HYPOTHETICAL CASE: TROOP B

Troop B of the Shiprock Highway Patrol is responsible for law enforcement on the highways in and around Gotham City. Although not the state capital, Gotham City and its environs account for 40 percent of the state's population and a preponderance of commuter traffic. Troop B's commander, a twenty-year veteran of the roads, has a strong sense of duty and even stronger feelings about what constitutes an appropriate service level to the citizens of Shiprock. For the first five years of his command, Troop B led the state in enforcement performance levels.

A recent report from the department's central office indicates that the performance of Troop B has dropped dramatically over the past eighteen months. For example, traffic collisions involving personal injuries and property damage are up 43 percent from the previous year. A disproportionate percentage of such accidents involve drivers who are driving under the influence (DUI) of alcohol. The accident trend is particularly disturbing to the commander because he believes that the solution to the problem is vigorous enforcement of DUI statutes.

The commander is equally disturbed by an increased number of citizen complaints against the members of Troop B. Some 95 percent of these complaints involve verbal abuse of citizens by officers. A review of personnel files indicates that in the previous eighteen months disciplinary actions involving suspension without pay have amounted to an excessive one hundred days for the troop of thirty officers. These suspensions have occurred over such issues as neglect of duty, rule infractions, and insubordination.

Morale in Troop B is low. The officers are disgruntled over a recent statewide study that recommended pay increases of 15 percent for supervisors and only 5 percent for troopers. Other morale and discipline problems stem from an increased tendency to transfer problem troopers in other parts of the state to Troop B. The commander feels that the latter trend is offset by a tendency of the upwardly mobile to seek assignment to Troop B.

In addition, a review of logs shows that some troopers take as much as two hours per shift for meals and breaks, which in itself would account for much of the decline in enforcement efficiency. Some troopers also take an inordinate amount of sick leave. State regulations allow employees to accumulate up to ten hours for sick leave per month; leave not used within forty-five days is lost. The troopers who exhaust all their available leave each month make it necessary to pay other troopers overtime to cover the zone. Finally, there has been bickering between troopers and dispatchers over call assignments.

The troop commander has resolved to take immediate steps to remedy the situation, but he sees that the problems may be beyond the expertise available in the department so he retains an outside consultant. Assume you are that consultant. What would you recommend to the commander? Specifically—

1. What sort of problem identification activities would you engage in?
2. What are the problems?
3. Assuming that your list of problems is substantially the same as the commander's, what solutions would you recommend?
4. What sort of implementation strategy do you recommend?
5. How do you propose to tell whether your solutions worked?
6. Suppose the commander was so impressed with your work in this matter that he engaged you to perform a full-blown evaluation of the troop and its mission. Develop an evaluation design to measure the agency's performance.

FOR FURTHER READING

Anthony, Robert N., and Herzlinger, Regina E. *Management control in non-profit organizations.* Homewood, Ill.: Richard D. Irwin, 1975.

Bingman, Charles F., and Sherwood, Frank D., eds. *Management improvement agenda for the eighties.* Charlottesville, Va.: U. S. Office of Personnel Management, Federal Executive Institute, FEI B-26, 1981.

Cook, Thomas D., and Reichardt, Charles S. *Qualitative and quantitative methods in evaluation research.* Beverly Hills, Calif.: Sage Publications, 1979.

Filstead, William J., ed. *Qualitative methodology.* Chicago: Markham, 1970.

Patton, Michael Quinn. *Creative evaluation.* Beverly Hills, Calif.: Sage Publications, 1981.

———. *Utilization-focused evaluation.* Beverly Hills, Calif.: Sage Publications, 1978.

Stone, Eugene. *Research methods in organization behavior.* Glenview, Ill.: Scott, Foresman, 1978.

Chapter 7

PUBLIC POLICY AND SOCIAL PROBLEMS

This chapter is a collection of related topics, each of which is an important dimension of evaluation research. These topics are normally treated at the beginning of a text as part of the justification for studying evaluation. In some cases, the topic is the subject of a complete text. We treat these topics last because a broad perspective on the field is of less importance than acquisition of basic evaluation skills. In addition, having acquired a knowledge of evaluation design and implementation, the student can better assess the uses of evaluation research for making and implementing social policy.

The chapter begins with a look at how government programs are influenced by the socioeconomic environment in which they operate and how well-intentioned social policies can have negative consequences for other programs and for society at large. This discussion establishes the groundwork for advocacy of a comprehensive application of the principles of program planning and evaluation to government programs. All topics are treated although some are merely sketched.

Such a comprehensive application of evaluation technology begins with the use of quantitative techniques to correctly identify social problems, followed by correct definition of the problem that the evaluation is to address and selection of a methodology appropriate to the needs of evaluation consumers as well as of the evaluators. The text then addresses the need for a national evaluation policy in order to maximize the benefits of evaluation research for managers who must implement or reject evaluation recommendations. Relevance for line managers can be achieved by evaluation designs that examine program structures and procedures as well as outcomes, an increased capacity for in-house evaluations, a national program to provide evaluation training for managers and executives as well as evaluators, and development of a national system for sharing evaluation designs and findings.

THE SOCIOECONOMIC ENVIRONMENT AND PUBLIC POLICY

Systematic evaluations of public programs have generated useful information about the nature of social problems and about government's ability to deal with them. As more and more programs were tried, adapted, and reapplied, definitions of problems had to be revised.

Observers learned, for example, that poverty went up even as the Gross National Product rose, unemployment shrank, and government programs proliferated.[1] Technological growth, worker displacement, urban migration patterns, and racism, as well as lack of money, can cause poverty.[2] As industry becomes increasingly automated, workers must be more highly skilled, and the same volume of production can be achieved by fewer workers. Displaced workers are relatively unskilled or have skills that are no longer valuable in the labor market.

When the economy is expanding, the decline in manufacturing jobs is offset somewhat by an expanding service industry.[3] Recently the economy has simultaneously experienced increased workers entering the work force (the result of the postwar baby boom), high inflation rates, and decreased productivity. The result is a sluggish economy that provides fewer opportunities to escape the cycle of poverty.[4]

Urban migration patterns have also contributed to the problems of poverty and of government efforts to deal with them. Traditionally, new immigrants arrived in America, established residence in the central city, then migrated outward. The acquisition of marketable job skills and a facility with the English language and American customs were prerequisites to outward migration. Southern Blacks, rather than White Europeans, were the post-World War II urban immigrants. Many were displaced by advances in agricultural technology, others merely migrated in search of a better life. Like their predecessors, these

---

[1] This point is made effectively by Gilbert Y. Steiner, *The state of welfare* (Washington, D.C.: Brookings Institution, 1971), p. 32. See also Harold L. Wilensky, *The welfare state and equity* (Berkeley: University of California Press, 1975), p. 32.

[2] Theodore Lowi, *End of liberalism* (New York: Norton, 1969), pp. 214–249.

[3] For a discussion of government efforts to manipulate the economy, see Arthur M. Okum, *The political economy of prosperity* (New York: Norton, 1969).

[4] See K. William Kapp, *The social costs of private enterprise* (New York: Schocken Books, 1971).

new urban immigrants possessed few skills, although technological advances in manufacturing meant there was less need for large numbers of unskilled workers. Because Blacks were also faced with racial stereotypes and outright bigotry on the part of many Americans, acquisition of work skills and middle-class values was not sufficient to allow them to migrate out of the central city.[5]

NEGATIVE EFFECTS OF PUBLIC POLICY

Government programs can also contribute to the complexity of social problems and obstruct their resolution. Federal housing policy, for example, contributed to urban decay and to segregation of neighborhoods. Public policy makers sought to stimulate the economy through the housing industry by such means as government efforts to underwrite mortgages through the Federal Housing Administration (FHA) and the Veterans Administration (VA) and deductions of mortgage interest on federal income tax forms. In addition, because the American dream is a home of one's own and because land was plentiful, the housing industry focused on housing tracts in outlying areas. As middle-class families took advantage of housing opportunities, they migrated to the suburbs, thus depriving the central cities of the tax base necessary to deal with the costs of education, welfare programs, crime, and so on.[6]

Housing policy contributed to segregation simply because many more Whites than Blacks were able to take advantage of the programs. Because minority neighborhoods were often redlined, federal housing loan guarantees were denied to many Blacks, and government programs previously underwrote loans in housing developments that were blatantly segregated. Furthermore, Blacks entitled to participate in FHA or VA programs could not purchase homes where they wanted to.[7]

Urban renewal policies also contributed to American social problems. Efforts to rejuvenate downtown areas by building municipal centers and shopping malls

---

[5] For an analysis of poverty among minorities and ethnic groups, see Daniel P. Moynihan, "Poverty in Cities," in *Urban studies,* ed. Louis Lowenstein (New York: Free Press, 1971), pp. 180–195.

[6] For a complete discussion of the problems of the central cities as they relate to suburban growth, see Frederick M. Wirt, Benjamin Walter, Francine F. Rabinovitz, and Deborah R. Hensler, *On the city's rim: Politics and policy in suburbia* (Lexington, Mass.: D. C. Heath, 1972). See also the discussion by the National Commission on Urban Problems, "The City as It Is and as It Might Be." in *Urban studies,* ed. Louis Lowenstein (New York: Free Press, 1977), pp. 100–101.

[7] For a complete discussion of federal housing policies and problems, see Gilbert Y. Steiner, *The state of welfare,* pp. 122–190.

did as much to destroy low-cost housing for the poor as they did to beautify city centers. These projects, moreover, have not succeeded in luring suburban shoppers back to the city.

These examples show how a program aimed at one problem can negatively spill over and contribute to another problem. We should note that public policy problems are rarely so simple that they can be corrected immediately with no second-order consequences, but the frequency and magnitude of spillovers could be reduced by placing a greater emphasis on the systematic study of social problems prior to committing resources.

THE NEED FOR A LARGER DATA BASE

To date, corrective programs have been put in place with little knowledge about the social problems they were to deal with, but advances in data-gathering technology now make it possible to create an unprecedented data base from which public programs can be designed and subsequently evaluated. Alice Rivlin, former head of the Congressional Budget Office, recommends establishment of a national social data bank that would make available data on such things as educational attainment, social adjustment, and consumption of public services, based on studies of persons from various backgrounds.[8] This information would increase our understanding of social problems and improve our ability to deal with them. At present we do not know, for example, what factors cause a person to break the cycle of poverty and become a productive citizen, or what causes persons from working- or middle-class backgrounds to become dependent on the welfare system. Until we have data on the causes of social problems, policy makers will continue to fund programs that are narrowly defined and/or that only partially address the problems at which they are aimed.

The complexities of planning social programs based on limited data can be illustrated by the example of education. Because acquiring education beyond high school is an excellent way to achieve upward mobility in the United States, educational programs should effectively apply public resources to help those at the lower level of the social system rise. The dilemma of the decision maker is where in the educational system to apply the resources. If the goal was to upgrade and improve educational opportunities for all citizens, then institutions of higher education might be the point at which to apply resources, but because aid to higher education is usually a subsidy for the middle class rather than the poor, decision makers devised a system of basic education grants and federal

[8] Alice M. Rivlin, *Systematic thinking for social action* (Washington, D.C.: Brookings Institution, 1971), pp. 9–46.

student loan programs to provide aid directly to economically disadvantaged students. Unfortunately, this aid goes only to those who manage to make their way through the primary and secondary education systems. Higher education grants cannot help those who do not acquire basic academic skills.

A national data base would provide information on those who do make it through the system and those who do not. It would also provide information on which special education programs are beneficial and which are ineffective consumers of public funds. The national data base should also provide clues to factors in the social environment that offset and nullify the gains made in the educational process. With all this knowledge, policy makers can begin to design programs for those who have fallen by the wayside.

Rivlin suggests that the knowledge gained from systematic problem analysis be used to make changes in *current* programs. Decision makers are increasingly unwilling to allocate resources for new programs, especially when the proposed programs are aimed at problems that old programs were supposed to solve. Furthermore, administrators of current programs are likely to resist new programs that will compete for scarce resources. Modification of current programs would be less threatening than wholesale changes.

USING EVALUATION TECHNOLOGY APPROPRIATELY

Whether or not decision makers decide to sponsor systematic assessments of social problems, evaluations of public programs will continue. Determining whether existing programs are accomplishing their missions is more important to policy makers than proper identification of the origins of social problems. The relative success of programs, however, depends on who is asking the questions and to what ends. Should policy makers, for example, accept evaluation findings at face value when those findings are based on limited studies? On the other hand, would funding of comprehensive evaluations be a cost-effective use of public resources?

Program administrators often view evaluation efforts as threats to their programs, but some resistance to evaluations can be traced to poorly designed and conducted evaluations to which administrators had previously been exposed, and some can be attributed to administrators' unfamiliarity with evaluation technology. In this section, we explore the related problems of design validity, the formulation of policy-relevant questions, and the problems of making evaluations relevant and useful to program administrators.

PROBLEMS OF DESIGN VALIDITY

A research design that is less than pure but that answers the appropriate questions is preferable to a design that is methodologically correct but useless to decision makers. The task of the evaluator is to meaningfully translate the goals of the program into a research design that produces useful data. The problem of generating a valid design is compounded by the reluctance of legislators and program staffs to help in the process.

For example, Congress rarely specifies program goals precisely. One evaluation team, whose task was to determine the level of inequality in U.S. schools, found no clear-cut congressional definition of educational inequality. The team had to decide whether educational inequality was to be measured as a function of community inputs to the schools, as the result of segregation, as a function of intangible variables such as teacher morale and teacher expectations of students, or as inequality of results given the same individual inputs–or whether inequality should be measured in terms of the consequences of segregated schools for persons of unequal backgrounds. It chose the fourth measure (inequality of results, given the same level of individual inputs), which also provided information on the other forms of inequality.[9]

A second problem faced by evaluators stems from the reluctance of administrators and program staffs to help in externally generated evaluations. Such evaluations are perceived as threats, as indications that somebody in a position of authority has questions about the program. The problem is compounded because externally generated evaluations frequently employ only output measures, which will tell whether a program is working but not why or why not. Why the program is *not* working is as relevant for the administrator as why it is working. Cooperation of the program staff can be improved by including measures that gather data relevant to the staff in addition to output measures. Staff resistance can also be reduced by seeking staff input in the developmental stages of the evaluation–before the research questions are firmly fixed.

FORMULATING POLICY-RELEVANT QUESTIONS

How research questions are formulated can affect whether a program is continued, expanded, or eliminated. One or more of the following questions usually guides evaluators in development of designs. Focusing on only one of the ques-

[9] James S. Coleman, "The Concept of Equality of Educational Opportunity," *Harvard Educational Review, 38,* 1 (Winter 1968), 7–22.

tions, however, may result in an incorrect or overly critical evaluation of an effective program.[10]

1. Does the program achieve its stated goals?
2. Does the program reasonably address the needs of its recipients?
3. How does the cost of the program compare with its benefits?
4. How does the program compare with alternatives?

Does the Program Achieve Its Stated Goals? The question of whether a program achieves its stated goals is central to any evaluation formulation, but Congress frequently states its intent in broad terms that are subject to a variety of interpretations. Because of this, it is not surprising that federal and state program officials sometimes differ about the intent of a program and the indicators by which the program should be measured.

Such differences can be illustrated with the vocational rehabilitation programs established to help handicapped persons gain marketable job skills. Rehabilitation clients may be people who are no longer able to continue in their chosen occupations because of accidental disability, or people with disabling physical or emotional illnesses, or people who are mentally retarded. For example, a construction worker who had extensive back injuries in an automobile accident might no longer be able to perform the lifting and bending chores associated with construction work. The rehabilitation agency would first assess that person's interests and talents, as well as his or her disabilities, to arrive at an occupational alternative. Then it would pay for necessary training and help the client find work in the new occupation. The assessment, training, and placement process is generally the same for all types of clients.

The success of such rehabilitation programs was traditionally measured in terms of the number of clients who had been placed in jobs. However, the definition of what constituted a disabled person who was able to benefit from the services varied considerably from state to state. Some states focused their efforts on persons with relatively minor disabilities, such as a factory worker who had lost one or two fingers. An agency might place this client in a new job that did not require a high degree of manual dexterity or a great deal of retraining. If the person was subsequently laid off, he or she could return to the agency for more training and placement and yet again be counted as a successful rehabilitation. A state agency could repeatedly place the mildly disabled while refusing services to extremely disabled persons, but because job placement was the

---

[10] This discussion draws heavily on Joseph S. Wholey, John W. Scanlon, Hugh G. Duffy, James Fukumoto, and Leona M. Vogt, *Federal evaluation policy: Analyzing the effects of public programs* (Washington, D.C.: Urban Institute, 1975), pp. 19–28.

measure of success, the agency could demonstrate that it was successfully achieving its stated goals.

Pressure from groups representing the disabled, and reformists in the rehabilitation field, produced a redefinition of standards. Federal guidelines now require that at least 50 percent of an agency's cases be severely disabled persons. The new standards have placed a strain on program resources because much greater effort is necessary to make the severely disabled work ready. Pressure from groups representing the disabled has also caused a redefinition of the rehabilitation function. Now disabled persons are demanding that agencies provide services to the severely disabled that will allow them to maintain independent residences whether or not they ultimately obtain employment. Even though state agencies are coming closer to accomplishing the stated goals of the program, the total productivity per dollar expended has noticeably declined.

Future evaluations of rehabilitation programs must therefore carefully define the measures used to determine program success in order to account for the redefinition of the rehabilitation function. If successful job placement is to continue as the principal measure of success, then placements of the severely disabled must be given extra weight, and repeated placements of the same client must be discounted. Furthermore, what constitutes a successful rehabilitation with regard to establishment of an independent residence for a severely disabled person must be considered in the development of the evaluation design.

Does the Program Reasonably Address the Needs of Its Recipients? The question of whether the program addresses the needs of its recipients is more difficult and costly to address than the question of whether the program is achieving its stated goals. The desire of severely handicapped persons to maintain their own residence is an example of how agencies have had to revise their mission in order to meet the needs of their clients.

Housing programs provide additional examples of the client-need test of program effectiveness. For example, housing officials may be able to demonstrate that public housing units operate at 95 to 100 percent occupancy, but public housing laws are designed to meet the needs of a variety of clientele. Senior citizens, for example, constitute a significant percentage of the housing poor, but they are reluctant to occupy public housing units that also service the needs of nuclear families. The elderly do not want the welfare stigma that is attached to public housing, and they fear the high crime rate that is commonplace in public housing.[11] In response, housing projects exclusively for the elderly have sprung up across the nation. Some housing authorities limit new public housing construction to projects exclusively for the elderly. The question then becomes

[11] Gilbert Y. Steiner, *The state of welfare,* pp. 155-158.

whether this emphasis on the elderly is at the expense of the equally deserving younger poor who are struggling to escape from poverty.

How Does the Cost of the Program Compare with Its Benefits? The quantifiability of cost-benefit ratios in terms of service levels per resources expended varies from program to program. Straightforward cost-benefit ratios are possible in instances where program outputs are readily quantifiable. For example, the number of troops that can be flown combat-ready halfway around the world is a straightforward time and materials question, and the cost of such a program can also be readily calculated. Decisions, therefore, can be based on the force readiness that is desired and the resources available.

Social programs frequently are not amenable to cost-benefit calculations. For example, calculating the cost of a prenatal care program for expectant mothers in Appalachia would be simple, but developing measurable program benefits would be more difficult; one cannot calculate the dollar value of a full-term baby over a baby who is born prematurely. Program evaluators would also find it difficult to illustrate that a higher percentage of full-term babies born to program participants was due solely to the program. Other long-term benefits of the program would prove equally hard to quantify, for example, the potential reduction in health problems over a lifetime for persons whose mothers participated in the program and the contributions that healthier persons would make in terms of work attendance and general social productivity. Such a program might not fare well when compared with programs that can demonstrate immediate quantifiable benefits.

How Does the Program Compare with Alternatives? Comparison evaluations can be useful for assessing the relative merits of current programs and new programs designed to replace them. For example, a reading program may have been judged effective under experimental conditions at a university, but before being implemented it should also be pilot tested under actual classroom conditions. Program participants could then be compared with peers who underwent traditional reading training. Pilot testing would also enable researchers to assess the ability of classroom teachers to apply the methodology. Finally, it might reveal unanticipated spillover effects of the program on such things as a child's acquisition of overall language skills and any impacts that the program might have on classroom discipline.

Caveats for Applying the Criteria. Each of the foregoing criteria-questions has been applied to one or another program, and each represents a valid policy question for certain types of programs. However, inappropriate application of these criteria, or applying them in isolation from other equally valid criteria, presents a danger. The appropriate combination of questions will vary from program to program. Failing to carefully select the policy questions to be addressed is to risk the unaltered continuation of programs that are not working or the cutting back of programs whose benefits outweigh their disadvantages.

MAKING EVALUATIONS RELEVANT TO PROGRAM MANAGERS

Evaluation research will not achieve its full potential until it becomes a decision-making tool that is routinely utilized by program managers at all levels of government. Much of the data currently generated by national evaluations is of little use to state and local administrators. Data gathered from a number of regions and assessed in the aggregate do not reflect the extremes of program successes and failures or the performances of individual programs. Variability of program success can be attributed to such factors as variations in service populations from region to region, variations in program delivery structures, and variations in the training, attitudes, and enthusiasm of program staffs in various locations. Program managers are therefore reluctant to use agency resources to analyze program outputs that will be lumped together with data from other organizations in ways that are meaningless to their own operations.

Four steps are essential to making evaluation relevant to program managers:

1. Federal guidelines must be expanded to require assessments of program delivery systems as well as of program outputs.
2. Delivery agencies must establish evaluation units.
3. A national program of evaluation training for administrators must be founded.
4. A system must be established to provide for sharing information about program delivery systems and the conduct of evaluations.

EVALUATION OF PROGRAM STRUCTURES AND PROCEDURES

Congress may not be enthusiastic about additional funding for programs that appear to be expending excessive resources for administration rather than for service delivery. To be relevant to administrators, however, evaluations must assess program structures and procedures as well as program outputs. Such a requirement would represent only an incremental change in federal guidelines, which already require state and local agencies to document how resources are expended, to report the number of clients served, and so on. Requiring that program operating procedures and structures also be documented would provide valuable information to federal officials as well as to state and local administrators.

Data on administrative structures would allow federal officials to make post hoc comparisons between successful and unsuccessful program units, and they could be used in policy-level decisions regarding program continuation. The data also would allow federal officials to generate uniform guidelines requiring states to emulate successful program structures.

Requiring state and local administrators to engage in introspective evaluations of program structures and procedures could also have direct positive effects for program management. Most managers did not design the administrative structures they manage, and most assume their positions after the program has been in operation for a number of years. In addition, most initial administrative structures have had new programs and projects tacked onto them. If state and local officials were required to assess administrative structures as well as program outputs, administrators might be persuaded to reorganize program components in accord with the agency's current mission. Reorganization, in turn, could result in increased program efficiency and effectiveness.

IN-HOUSE EVALUATION UNITS

As program administrators come to appreciate the value of evaluation as a decision-making tool, the number of agencies that maintain their own evaluation capacity in the form of in-house evaluation units will increase. Evaluation units are advantageous because a certain amount of expertise is necessary to interpret and implement evaluation guidelines, and because they are more rigorous in their approach to evaluation than line managers who are assigned evaluation in addition to other duties. When a person sees his or her role as that of a line manager or a social work professional, for example, that person is likely to resist efforts to make him or her an evaluator as well. It must also be remembered that the line manager's day-to-day responsibilities will usually take precedence over the need to conduct a timely and valid evaluation. Evaluations that are conducted in-house, as opposed to externally conducted evaluations, usually emphasize program-specific problems of extreme interest to the agency administrator and meet with less resistance from program personnel.

In-house evaluation units also can provide program administrators and managers with hard data on a variety of subjects even when they are not actually engaged in evaluations. These services might include needs assessments of clientele groups, the preparation of policy and procedure manuals for the agency, and assisting the administrator in his or her preparation of the agency's legislative package from year to year. In short, program planning and evaluation units can provide the agency with a full range of staff services.

A NATIONAL EVALUATION TRAINING PROGRAM

Despite the growing reliance of public officials on evaluation data for making program decisions, the majority of line managers simply do not have an adequate understanding of the technology and uses of evaluation research. Without this understanding, even agencies that do have evaluation units cannot reap the full benefits of evaluation technology. Line managers, for example, might not seek assistance from the evaluation unit simply because they are unaware of the types of data that can be generated. There is also the potential for development of line-staff frictions between the evaluation staff and the more traditional staff functions, such as personnel and accounting. For example, line personnel may be reluctant to contribute their valuable time to what they see as a vague staff project that is of interest only to the federal government or the boss. In short, in-house evaluation findings may be ignored, resisted, or sabotaged as readily as those of externally conducted evaluations.

A logical first step in demystifying evaluation technology would be a training program to acquaint top administrators with the basic concepts of evaluation design and the uses of quantitative data in executive decision making. One approach would be for the federal Office of Personnel Management (OPM) to establish a training program for top administrators to deal with the concepts of evaluation in a general manner. Federal oversight agencies could then supplement the OPM course with program-specific training.

Among the obstacles to this strategy is the possibility that administrators who do not have data-gathering and analytical skills may be reluctant to acquire them out of fear of the unknown. Others may think they cannot take the time for training. To succeed, the training program would have to focus on the basic concepts of evaluation and how it can be useful to managers. To minimize the amount of time administrators had to spend away from the agency, courses could be given in an intensive week-long format. Once top management understood and saw the benefits of evaluation technology, training could be directed at mid-level managers.

The goal of training is not to make all program administrators and managers evaluation experts. Instead, it is to make them effective consumers of evaluation data, able to understand what types of data are appropriate to an impending decision and/or to a program-specific evaluation. An effective consumer also understands when an impending decision warrants the expense and delay involved in extensive data-gathering efforts. Finally, an effective consumer is sensitive to the problems of the evaluator and the latter's need to have the cooperation of line managers and program staff in conducting an evaluation.

A NATIONAL SYSTEM FOR SHARING EVALUATION INFORMATION

The final component necessary to make program evaluation an effective tool of management is a national system for sharing evaluation information. At present, because there is no mechanism for sharing information among evaluators or program administrators, a successful program format or an inexpensive evaluation design may go unnoticed by persons outside the specific agency.

The Urban Institute has called for establishment of an evaluation clearinghouse for cataloging and sharing evaluation information. The clearinghouse approach could be a valuable resource for program administrators and their staffs in developing appropriate evaluation strategies. The clearinghouse also could collect and disseminate data on various administrative structures and their relative successes so that administrators could compare their operating procedures and program outputs with those of comparable agencies. Such comparisons could result in considerable resource savings to agencies that otherwise might experiment with delivery formats that had not worked elsewhere.

The diversity of programs now undergoing evaluation may limit the utility of a single, comprehensive clearinghouse. Collecting and cataloging information on a variety of programs would require a large and expensive organization, and service to specific categories of consumers might suffer from the effort to offer comprehensive services. As an alternative, evaluation clearinghouses could be operated on a program-specific basis. The U.S. Department of Education, for example, could provide clearinghouse services to educational institutions and state education agencies. Consortiums, such as the League of Cities, in the various states also could provide clearinghouse services to their members on a more limited basis.

To ensure maximum participation, clearinghouse data must be cataloged and stored without identifying the contributing agency, so that program administrators would not be reluctant to share information that reflected badly on their programs. If anonymity was not assured, evaluation clearinghouses could become repositories only of evaluations that produced positive findings. Maximum participation could be further assured by making clearinghouse participation a requirement in cases where the evaluation was mandated as a condition of federal funding.

SUMMARY

The evaluation of public programs has undergone exponential growth in the past two decades, but evaluation science has not yet achieved its full potential as a decision aid in all phases of the policy process from policy formulation to administrative implementation. At the formulation stage, it is now possible to generate comprehensive data on the nature and origins of social problems. Evaluation research can be valuable in helping decision makers assess the efficiency of various programs. Policy makers must realize, however, that evaluation research is a value-neutral technology that can be applied for good or for ill, depending on the research questions the evaluation is designed to address. Evaluation technology must therefore be applied in an evenhanded fashion across programs.

The problems of design irrelevance and administrative resistance that have plagued previous evaluation efforts can be overcome by showing managers and their staffs how they can make evaluation work for them. This could be achieved, in part, by the establishment of in-house evaluation units that could strengthen the relevance of evaluation designs to program needs and lower the resistance of line personnel to the evaluation findings.

The greatest potential of evaluation technology is its use as a decision-making tool in the hands of program managers. An important first step in achieving this potential is establishment of a national program of managerial education to instruct administrators at all levels of government in the uses of evaluation data as a decision-making tool. The proliferation of evaluation studies makes possible the establishment of a system of evaluation clearinghouses and consortiums, a system that would provide a valuable resource for collecting and disseminating information about evaluation designs and successful programs and structures.

FOR FURTHER READING

Cronbach, Lee J., Ambron, Sueann Robinson, Dorabusch, Sanford M., Hess, Robert D., Hornick, Robert C., Phillips, D. C., Waller, Decker F., and Weiner, Stephen S., *Toward reform of program evaluation.* San Francisco: Jossey-Bass, 1980.

Horowitz, Irving, and Katz, James. *Social science and public policy in the United States.* New York: Praeger, 1975.

Rivlin, Alice M. *Systematic thinking for social action.* Washington, D.C.: Brookings Institution, 1971.

Weiss, Carol H., ed. *Using social research in public policy making.* Lexington, Mass.: D. C. Heath, 1977.

Wildavsky, Aaron B. *Speaking truth to power: The art and craft of policy analysis.* Boston: Little, Brown, 1979.

Wholey, Joseph S., Scanlon, John W., Duffy, Hugh G., Fukumoto, James, and Vogt, Leona M. *Federal evaluation policy.* Washington, D.C.: Urban Institute, 1970.

Wholey, Joseph S. *Evaluation: Promise and performance.* Washington, D.C.: Urban Institute, 1979.

Glossary

Audit: An examination of a program or project to verify that its implementation meets stated objectives and rules or regulations.

Causal modeling: Constructing a theoretical process that "explains" events by identifying the direction of the change one phenomenon has on another and that can be tested with empirical data.

Congressional Budget Office: Created by the U.S. Congress in 1974 as part of an effort to improve its budget actions, the CBO provides staff support and computer-based analysis for the House and Senate members to help them anticipate and monitor the broad economic impacts of specific legislation.

Correlational analysis: Systematic examination of certain phenomena, usually expressed as data, to determine whether the phenomena are related or associated in the sense that when one phenomenon occurs the other also occurs.

Cost-benefit analysis: An analytical framework for evaluating a program or project through a comprehensive assessment of both the advantages (profits) and disadvantages (costs) that would occur if the program were implemented. Applications range from a broad, conceptual evaluation to a quantitatively oriented comparison of specific projects.

Cost-effectiveness analysis: A type of cost-benefit analysis in which either the costs of the programs being evaluated are identical so that it is necessary to compare only the benefits, or the benefits of the programs are identical so that only the costs need to be compared.

Delphi technique: Developed at the RAND Corporation, the Delphi technique attempts to improve the judgments of a group of experts by providing them with anonymity while they are asked questions about problems and issues. The group is then supplied with a summary of the results and asked to reconsider their original answers and justify or explain them if they do not move toward the consensus position that emerges.

Entropy: The second law of thermodynamics (Entropy Law) states that matter and energy can be changed only in one direction–from usable to unusable, or from order to disorder, or from available to unavailable, or from structure to chaos.

Equity: A desirable distribution of goods and services to all members of a society through a just allocation of costs and benefits.

Experimental design: A research model that utilizes random selection from the population of interest, control over the experimental variable or event, and a control group for comparing the performance or results with the experimental group.

Experimental mortality: The loss of subjects from an experiment due to such factors as illness, lack of interest, or refusal to participate, which affects the comparability of results between the experimental and control groups.

Externalities: The unintended and often unrecognized costs and benefits of a project which accrue to those outside the scope of a cost-benefit analysis. These costs and benefits are typically difficult to value in monetary units.

External validity: The extent to which measurements obtained in a particular situation or context can be generalized accurately to other situations or populations of interest.

Federal Register: Since 1936, the public record of every regulation and related amendments having legal effect that have been issued by federal agencies with congressional or presidential authority to do so. The regulations and amendments for the preceding year are codified every January 1 into the *Code of Federal Regulations.*

Genius forecasting: In contrast to highly computerized forecasting models, using experts to "brainstorm" how future events will occur in order to take advantage of the experts' knowledge, creativity, and intuition.

Goal reification: Defining the goals of an ongoing program specifically for the purposes of designing an evaluation. The danger of goal reification is that the definitions can become so abstract that they make the subsequent findings of the evaluation meaningless to program officials.

Gross National Product (GNP): A measure of the goods and services produced by the United States, aggregating personal consumption expenditures, gross private domestic investment, net exports of goods and services, and government purchases of goods and services.

History: Extraneous events, outside the control of a study but occurring between the first measurements and subsequent measures, which may have affected the results of the study.

Incommensurables: Costs and benefits stemming from completion of a project which appear to fall outside the typical measures used in the analysis although they may be measurable in another context.

Instrumentation: A loss in the effectiveness or consistency of the measuring instrument from, for example, change in the springs of a scale or changes in the observers or methods of scoring.

Intangible effects: A term used in cost-benefit analysis to designate consequences that appear to be beyond quantitative or economic expression.

Interaction effect: The effect of a combination of the threats to validity, which may be more significant than the individual threats.

Internal rate of return: The discount rate that would reduce the present value of a project under consideration to zero.

Internal validity: The extent to which an indicator or set of indicators accurately measures the concept or variable in the situation or population of interest in a particular study.

Market pricing: Pricing or valuation based on the interaction of buyers and sellers as they exchange goods without artificial or arbitrary restrictions.

Matched-pairs designs: Designs in which individuals or units under study are assigned in pairs to control and experimental groups according to characteristics that they hold in common. For example, one might match fourth-grade boys on the basis of reading scores.

Maturation: Any natural change in subjects during a study period, such as aging or learning, which was not caused by the study itself but may have affected the results.

Multiple regression: A quantitative approach used to determine how well two or more events together seem to predict the occurrence of another event, either singly through use of statistical controls or together.

Multiple treatment interference: The effect of applying multiple treatments to the same respondents, thus precluding an assessment of the impact of any one treatment.

Net present value: A number representing the current value of a potential project after the stream of costs and benefits expected to occur over the project life are discounted and aggregated.

Nonequivalent control groups: When random assignment to groups is not possible, individuals or groups with characteristics similar to those of the participants in the program or experiment are used as controls.

Nonexperimental designs: Research designs that do not account for or control for alternative explanations for the change that may occur, for example, before-and-after studies of a program or an after-only study. These are particularly useful for descriptive purposes and for obtaining insights about on-going programs.

Office of Management and Budget: Formerly the Bureau of the Budget, the OMB is the division of the Executive Office of the President with primary responsibility for preparing the executive budget presented to Congress.

Participant-observer method: An evaluation method in which observers take an active part in the situation under study in order to record events with more insight.

Pecuniary benefits and costs: Terms used in cost-benefit analysis to distinguish external and financially related effects stemming from completion of a particular project, such as an increase or decrease in the cost of goods or services that depends on the project output.

Program validity: An evaluation design that has program validity utilizes indicators to test program outputs and statistical manipulations that are understandable to program officials. The reports of such evaluations should provide information that is useful in upgrading program performance.

Project calendar: A device used to plot the time frame in which various components of a project are to be completed; it may also illustrate the points at which various expenditures must be met.

Random selection: A sampling process through which individuals in the population of interest have an equal chance of being chosen and in which the choice of one individual has no effect on the choice of any other individual.

Responsibility matrix: A management control tool whereby the duties of specific individuals in carrying out planned changes are juxtaposed against the various components of the change strategy.

Risk-benefit analysis: A type of cost-benefit analysis in which the negative consequences of a project or program are measured in terms of the types and magnitude of risks to individuals or to communities instead of in monetary units.

Selection bias: Occurs when random sampling or matching for ensuring comparable experimental and control groups is not feasible and subjects are chosen in a manner that might affect the results.

Sensitivity analysis: A slight varying of the values of parameters or variables in a model in order to see the effect of such changes on the outcome. Particularly useful when there is uncertainty about the accuracy of the data used for the analysis.

Shadow pricing: Approximate valuations of true values for goods and services used by analysts when there is no market for the good or service, a situation which exists for many public projects. Shadow pricing is accomplished by substituting the value of a similar good for the actual good being considered, for example, substituting the cost of renting an hour's use of a private tennis court for the value of an hour's use of a court in a public park.

Solomon four-group design: An experimental design that addresses the problem of external validity by controlling both for testing effects and for interaction effects between testing and treatment. This design uses the pretest-posttest with and without treatment and the posttest only with and without treatment.

Statistical regression: If study subjects are initially selected because of extreme scores, their scores on the subsequent measures can be expected to move inward toward the mean score. This is known as statistical regression.

Structural functionalism: Talcott Parsons' theoretical framework for approaching problem solving, which suggests that structures (hierarchies, social classes, departmentalization) can best be understood and explained by looking at the functions they serve in the broader context.

Suboptimization: Utilization or management of resources in a less than optimum or less than efficient way.

Tangible effects: A term used in cost-benefit analysis to designate consequences that can be identified and measured in some fashion.

Testing effects: The effects that the initial act of testing or measuring a subject's performance can have on the subsequent tests or measures.

Time series design: The extension of the basic pretest-posttest design to repeated measures of a particular set of variables or individuals, usually characterized by a specific time interval.

Welfare economics: A branch of economics concerned with providing a method of evaluating the social gain or loss stemming from economic changes. Theories of welfare economics have been constructed using individual preferences and social preferences.

Index

Accountability subsystem, 8-9
Adaptation subsystem, 7
Advocacy planning, 29-30
Agency planning, 31-32
Applying systems theory, 9

Balkinwalk, Virginia case study, 155-157
Beckhard, Fred, 26, 49
Bell, Cecil H., 110
Benefits and problems of experimental designs, 92-93
Block, Deter B., 120
Botner, Stanley, 30
Bozeman, Barry, 23, 109

Campbell, Donald, 81, 86
Change of Command case study, 157-159
Clayton, Ross, 37-38
Clotfelter, Charles T., 70
Coleman, James, 107, 166
Community health example, 114
Comparing programs and alternatives, 170
Consensus building, 109-111
Cost-benefit analysis, 48-78
  benefit-cost ratios, 48, 66-67
  classification of cost benefit data, 53
  common units of measure in, 50
  comparing costs and benefits, 65-67
  consumer surplus, 57-58
  consumption method, 61-62
  cost effectiveness, 48
  cut-off period, 66
  dangers of, 170
  equity in, 67
  impacts, 50
  incommensurables, 54, 60-62
  internal and external effects of, 53-54
  internal rate of return, 66
  pros and cons of, 69-70
  real v. pecuniary benefits and costs, 54
  risk benefit analysis, 48
Cost-benefit analysis (*continued*)
  sensitivity analysis, 67
  tangible and intangible effects, 54
  techniques of cost-benefit analysis, 51-55
Cost-Benefits of the 55 mph Speed Limit cases, 70-78
Creaming, 100
Critical Path Method, 34
Cutback Management case, 131-134

Defining program goals, 108-111
Delphi technique, 109-110
Demand curves, 56
Design selection, 89-102
Design validity, 79-89, 167
  external validity, 79-80
  internal validity, 80-81
Despot suppression syndrome, 140-141
Determining goals, 21-22
Developing and assessing alternatives, 25-26
Dickson, W. J., 83
Discount rates, 62-65
Dolbeare, Kenneth M., 107
Dusbow case study, 42-45

Elazar, Daniel, 3, 10
Evaluating program structures, 171-172
Evaluation clearinghouse, 174
Experimental designs, 89-95
Experimental police program, 120-123
Experimentation case study, 123-125
Externalities, 59

Feedback phase of planned change, 154-155
Forecasting, 22-25
  correlation analysis, 24
  Delphi, 23
  methods of, 23
  trend analysis, 23-24
French, Wendell L., 110

Group problem solving techniques, 110-111

Hahn, John C., 70
Harris, Ruben T., 26, 149
Hawthorne effect, 83
Hidden agendas in evaluation, 106-108

Implementation of changes, 146-154
Indications, 15-16
In-house evaluation units, 15-16
Internal v. external evaluations, 104

Johnson, Ronald W., 107

Kapp, William K., 163
Kendall, Patricia, 92

Lave, Charles A., 74
Legislative intent, 111-113
Little, I. M. D., 61
Lowi, Theodore, 3, 163

Maintenance subsystem, 8, 13-14
Managerial audit, 142
Managerial subsystem, 7-8
March, James, 20
Marketing subsystem, 9
Market prices, 56
  absence of, 59-60
  biased, 57-59
Matched pairs designs, 99-100
Measures, 113, 116-119, 121-122
Meier, Kenneth J., 49
Migrant Workers case study, 126-127
Miller, William, 93
Mixed designs, 101
Monetary estimates, 55
Moynihan, Daniel P., 164
Multiple time series designs, 96-97

National Environmental Protection Act of 1969, 49
National evaluation training program, 173-174
National planning, 30-31
National social data base, 165-166
Negative effects of public policy, 164-165
Net present values, 65-66

Nonequivalent control groups, 97-99
Nonexperimental designs, 95-102

Office of Personnel Management, 104-105
Okum, Arthur, 163
Operational planning, 31-32
Operation neighborhood, 120
Organization environments, 2-5, 10-13
Organization goals, 6, 15-16, 21-22
Outcome evaluation model, 111-114, 122
Outcome valence, 114

Participant observer, 100-101, 141-142
Patton, Michael, 104, 135
PERT/CPM, 32-38
  critical path, 34
  events and activities, 33
  perturbations, 36-37
  slack, 33
  steps in, 37
  time estimates, 33-34
Pigou, A. C., 63-64
Planning, 19-30
  constraints on, 27-30
  implementation, 26-27
  principles and purposes, 19-20
  steps in planning, 20
Posttest only designs, 91-92
Powers, Edwin, 93
Preplanning evaluation, 104
Present value, 62-63
Pretest posttest designs, 89-91
Pricing techniques, 55-65
Problem identification phase, 138-142
  group interviews, 138-141
  personal interview, 108-109
  survey techniques, 138-139
Process approaches, 137-138
Process evaluation model, 137
Process evaluation overview, 155
Process outcome dichotomy, 135-136
Program elements, 113, 114-116, 121
Program goals, 111-113, 114, 121, 165-166
Program outcomes, 113, 119-120, 122-123

Program planning, 32
Program relevant evaluations, 171-174
Program structures, 39
Project calendar, 151-154
  example of, 152
Proportionate risk, 61
Proximate indicator, 94, 97, 113, 116, 121
Public goods, 59-60

Rapp, Donald, 76
Recipient needs criterion, 169-170
Responsibility matrix, 150-151
Roethlisberger, F. J., 83
Ryan, William, 141

Shadow prices, 57
Simon, Herbert, 20, 26
Social discount rate, 63
Solomon four, 91
Solution development, 144-146
Specht, David I., 120
Squeaky wheel syndrome, 139-140
Stanley, Jullian, 81, 86
Steiner, Gilbert Y., 163, 164, 169
Stokey, Edith, 50
Strategic planning, 31
Sylvia, Ronald D., 109
Systems theory, 5-20
  advantages and disadvantages, 16-17
  feedback, 6
  inputs and outputs, 5-6, 14-15
  suboptimization of goals, 9, 20
  subsystems, 7-9

Task force approach, 147-149
Task scheduling and reporting systems, 149-154
Teachers' Aide case study, 127-131
Technical subsystem, 8
Theoretical goals, 111-113, 114, 120-121
Thompson, Mark, 62
Threats to validity, 81-89
  effects of pretesting, 87
  experimental mortality, 85
  history, 81-82
  instrumentation, 83-84
  interaction effect, 86-87
  maturation, 82
  multiple treatment interference, 88-89
  reactive effects of experimental environment, 87-88
  reactive interactive effects, 83
  selection bias, 84-85
  statistical regression, 84
  testing, 82-83
Time series designs, 95-96, 118, 121
Training Evaluation case study, 125-126
Troop B case study, 159-160

Using evaluation technology, 166-171

Values of planners, 29
Van Gundy, Andrew B., 25
Vasu, Michael L., 29

Weiss, Carol H., 103, 106, 107
Welfare economics, 55
Wetlands case study, 41-42
Wholey, Joseph, 168
Wildavsky, Aaron, 31
Wirt, Frederick, 164
Witner, Helen, 93

Zeckhanser, Richard, 50